Mathematics and Physics in Classical Islam

Crossroads - History of Interactions across the Silk Routes

Edited by

Angela Schottenhammer
(*Catholic University of Leuven, Belgium*)

VOLUME 5

The titles published in this series are listed at *brill.com/cros*

Mathematics and Physics in Classical Islam

Comparative Perspectives in the History and the Philosophy of Science

Edited by

Giovanna Lelli

BRILL

LEIDEN | BOSTON

Drawing by Giovanna Lelli, freely inspired by Jalāl al-Dīn Rūmī' (1207–1273)'s story "The Elephant in the Dark House", Mathnavī Maʿnavī (Spiritual Poem), Book III.

Library of Congress Cataloging-in-Publication Data

Names: Lelli, Giovanna, editor.
Title: Mathematics and physics in classical Islam : comparative perspectives in the history and the philosophy of science / edited by Giovanna Lelli.
Description: Leiden ; Boston : Brill, 2022. | Series: Crossroads - history of interactions across the silk routes, 2589-885X ; volume 5 | Includes bibliographical references and index.
Identifiers: LCCN 2022001519 (print) | LCCN 2022001520 (ebook) | ISBN 9789004513143 (hardback : acid-free paper) | ISBN 9789004513402 (ebook)
Subjects: LCSH: Science–Philosophy–Islam–History | Islam and science–History
Classification: LCC Q174.8 .M375 2022 (print) | LCC Q174.8 (ebook) | DDC 509/.02–dc23/eng20220419
LC record available at https://lccn.loc.gov/2022001519
LC ebook record available at https://lccn.loc.gov/2022001520

Typeface for the Latin, Greek, and Cyrillic scripts: "Brill". See and download: brill.com/brill-typeface.

ISSN 2589-885X
ISBN 978-90-04-51314-3 (hardback)
ISBN 978-90-04-51340-2 (e-book)

This book is printed on acid-free paper and produced in a sustainable manner.

A well-known story of the Mathnavī Maʿnavī (Spiritual Poem) of the Persian mystic Jalāl al-Dîn Rūmī is known as "The Elephant in the Dark House". The story tells that people brought an elephant in a dark house. Many visitors came to see it. But because of the darkness they could only feel it with the palm. For one the elephant's trunk was like a water-pipe, for another its ear was like a fan, for another its leg was like a pillar, for another its back was like a throne. None could feel the elephant's real shape.

Contents

Illustrations

Notes on Contributors

Roshdi Rashed

Emeritus Research Director (distinguished class) at the CNRS (France), is a mathematician, a philosopher and a historian of science. Among his latest publications on the history of Arab science, *Ibn al-Haytham. L'émergence de la modernité classique* (2021).

Hossein Masoumi Hamedani

PhD (1994), Université de Paris VII, is retired professor of history and philosophy of science at the Iranian Institute of Philosophy, member of the Iranian Academy of Persian Language and Literature. His fields of research are the history of optics and the mathematical sciences in the Islamic period, history of philosophy and literary criticism. Among his latest publications : *Roshdi Rashed: Initiation à sa pensée et son œuvre* (2021).

Anas Ghrab

Ph.D. (2009), Université Paris 4-Sorbonne, is assistant professor of Musicology at the University of Sousse (Tunisia). His field of research is Arab musicology. He leads the project Saramusik.org (cataloguing and referencing Arabic manuscripts on music). His PhD thesis was about "An Anonymous Commentary of the *Kitāb al-adwār* : Critical Edition, Translation and Presentation of Arabic Readings of the Work of Ṣafī al-Dīn al-Urmawī".

Zeinab Karimian

Ph.D. in the history of science (University of Paris, 2019), Postdoc researcher at Sharif University of Technology. She works specifically on the History of mathematics in Islamic medieval world. Her Ph.D. thesis is about *The Recension of the* Conics *of Apollonius* by Naṣīr al-Dīn al-Ṭūsī: text, translation and commentaries of Book I.

Marouane ben Miled

PhD (2003), is a researcher at the Ecole nationale d'Ingénieurs (Tunis), and associated researcher at the CNRS (Aix-Marseille). His field of research is the history of Arabic algebra. He has published *Opérer sur le Continu* (traditions arabes du Livre X des *Eléments* d'Euclide), 2005.

Patricia Radelet-de Grave

PhD (1981), UCL Louvain, is a physicist and teaches history of mathematics and physics to mathematicians, physicists, engineers, architects and computer scientists at that university. Her main interests are the relations between mathematics and physics. Among her publications, *Entre mécanique et architecture* (with E. Benvenuto, 1995).

Paul D. Buell

(Ph.D. University of Washington, 1977) is adjunct professor at the University of North Georgia. He is an historian of Chinese medicine, of Food, Central Eurasia and the Mongol Empire. He is author or coauthor of many books and articles, including *Arabic Medicine in China* with Eugene Anderson.

Introduction

Giovanna Lelli

This volume is the result of a teaching and research project on the history and the philosophy of science of classical Islam initiated at the University of Gent (Belgium) in 2015. It gathers some of the contributions to a series of symposia organised in the same university in honour of Roshdi Rashed as a part of this project.[1] To these contributions, reworked in view of this publication, new ones have been added, which complete and enrich the picture.

With this volume we wish to contribute to the studies on the history and the philosophy of science which consider the development of modern science in Europe in the light of scientific advances achieved earlier in the Arab-Islamic world. From the nineth century, thanks to a massive translation movement, the Arab-Islamic world appropriated the Hellenistic scientific heritage. The advances accomplished by the Arabic sciences varied from one discipline to another and from one branch to another within the same discipline. This volume, which focuses on the mathematical and the physical sciences, highlights some of these advances, mentions concrete examples of them and indicates their general epistemological scope.

Rashed, in his "Science in Islam and Classical Modernity", examines how the knowledge of Arabic science[2] allows a better understanding of "classical science". The latter is traditionally considered to be the early modern European science which gradually replaced Aristotelian physics and cosmology with a new rationality characterised by mechanism, mathematisation and experiment. Rashed argues that the new rationality of classical science was introduced earlier by Arabic science between the ninth and the twelfth centuries. This new rationality was both algebraic and experimental. Thanks to al-Khawārizmī (780–850), algebra became an autonomous discipline. Eventually, the generality of its scope allowed the application of various mathematical disciplines to one another: arithmetic to algebra, algebra to arithmetic, both to

1 1st Symposium: "Islam and Science: a Living Heritage. Achievements in Classical Islam and Contemporary Challenges of World Science" (Ghent University, May the 8th 2015). 2nd Symposium: "The Arab-Islamic sciences (9th to 14th centuries) in the *longue durée*. The Scientific Revolution in question" (Ghent University, February the 12th 2016). 3rd Symposium: "Islam and Rationalism. The Development of the Scientific Method" (Ghent University, April the 28th 2017).

2 The science of classical Islam was written mainly in Arabic. Therefore, we use indifferently the term "Arabic science", "Arab-Islamic science" or the sciences of classical Islam.

DOI:10.1163/9789004513402_002

trigonometry, algebra to the Euclidean theory of numbers, algebra to geometry, geometry to algebra. Rashed mentions effects of these various applications: the development of polynomial algebra, combinatorial analysis, numerical analysis, the numerical resolution of equations, the new theory of numbers, the geometrical constructions of equations and the separation between rational and integer Diophantine analysis. In rational Diophantine analysis an explicit distinction was made between determinate and indeterminate problems, while in integer Diophantine analysis the infinity of solutions, approximate solutions and the impossibility of solutions were all considered true solutions. These features would be those of Diophantine analysis as understood by Bachet de Mézirac and Fermat in the seventeenth century, after a new appropriation of Hellenistic and Arab mathematics by Europe had begun in the twelfth century through translations into Latin, Hebrew and Italian. As far as the new experimental rationality is concerned, Rashed shows that the work of Ibn al-Haytham (965–1040)[3] is indispensable for understanding the emergence of a new experimental method in late seventeenth century physics in Europe.

Another feature of the sciences of classical Islam, which was tightly related to the new algebraic and experimental rationality, was the emergence of a new relationship between mathematics and physics. Classical Islam inherited the Aristotelian classification of the sciences based on the hierarchical tripartitions of the theoretical sciences into metaphysics, mathematics and physics. Mathematics, for its abstraction from sensual reality, occupied a higher rank than physics, and included arithmetic, geometry, optics, harmonics and astronomy. Physics dealt mainly with changing entities and studied the laws of motion of inanimate bodies, including the celestial ones. It also included biology and psychology.

Hossein Masoumi-Hamedani, in his "Physics and the Mathematical Sciences in the Islamic Period. A Conceptual and Historical Survey", shows that not all the Arab-Islamic mathematicians and philosophers agreed with the Aristotelian division of labour. The existence of problems studied both in mathematics and physics is already attested in Aristotle (e.g. in the case of optics). There were scholars in Islam who argued that mathematics plays a fundamental role in philosophical demonstrations or that mathematics can solve problems that cannot be solved by physics alone. This was the case of Ibn al-Haytham, whose optics included experiments and a combination of mathematics and physics, as it is attested to not only in his great book of *Optics*, but also in other specific

3 Throughout this volume, the dates of dynasties and the births and deaths of historical figures are often approximate when referred to pre-modern times.

works like a group of treatises on the light of the moon. Masoumi-Hamedani concludes by saying that the problematic nature of the relationship between mathematics and physics in classical Islam allows a better understanding not only of the history of science in Islam, but also of the history of early modern science (i.e. classical science).

In his article "Ibn al-Haytham: between Mathematics and Physics", Rashed explains, in a more detailed manner, the meaning of this new combination between mathematics and physics. In astronomy, Ibn al-Haytham, having found contradictions in Ptolemy, established a totally geometrical celestial kinematics, independent of cosmological considerations and of Aristotelian dynamics. The result was a model of the apparent motion of the "seven planets" halfway between Ptolemy and Kepler. In optics, Ibn al-Haytham reformed the optics of Euclid and Ptolemy, which was a geometry of perception, and modified the doctrine of the Islamic Aristotelian philosophers of Islam, who considered the forms perceived by the eye as "totalities" transmitted by the objects under the effect of light. He separated the theory of vision from the theory of light and established experimentally that light propagates independently of vision from illuminated objects onto the eye in straight lines and, he assumed, with great speed. In so doing, he founded a totally geometrical optics. The advances he accomplished in astronomy and optics were similar: he mathematised these disciplines and combined this mathematisation with the ideas of the physical phenomena.

Anas Ghrab, in his "La musique parmi les sciences dans les textes arabes médiévaux", deals with the position occupied by music in the Arab-Islamic system of knowledge. Before acquiring the status of an autonomous discipline, music was considered by most of the Arab authors as part of the mathematical sciences, according to the Aristotelian pattern. Aristotelian philosophers like al-Fārābī (878–950) and Avicenna (980–1037), relying upon Euclid and Ptolemy, observed that music shared elements with physics, because of the physical nature of sounds. In this respect Ghrab highlights another aspect of the relationship between different branches of mathematics and between mathematics and physics in classical Islam. He also refers to the impact of classical Islam in later research in music by the European scholarship of the Renaissance and beyond.

Zeinab Karimian, in her "Traditional and Modern Science in an Age of Transition: ʿAlī Muḥammad Iṣfahānī and the Logarithm of Numbers", analyses the case of an Iranian mathematician who, after having received an education in traditional mathematics, studied modern European mathematics and sciences at the Dār al-Funūn, a polytechnic founded by the Qajar dynasty in 1851. Al-Iṣfahānī, who introduced a more sophisticated method of interpolation in the calculation of logarithms, is an example of how classical mathematics was still able to dialogue with modern mathematics in the nineteenth century. It was an age of transition. Eventually, in the course of the twentieth century, European modern mathematics would be appropriated by the entire world.

If the writings of Rashed and Masoumi-Hamedani highlight relevant developments and advances of mathematics and physics in classical Islam with respect to the Hellenistic age, the writings of Marouane ben Miled, Patricia Radelet-de Grave and ours refer more specifically to the interdisciplinary openness of the Arab-Islamic culture as a factor of creativity and progress in mathematics, physics and the social sciences. Of course, interdisciplinary openness was not a specific feature of classical Islam. On the contrary, the tendency to hyper-specialisation and to a growing gap even between different branches of the same disciplines is a very recent feature of the culture of the twentieth century.

In his "Formalism and Language in the Beginning of Arabic Algebra", ben Miled deals with the influence of grammar and lexicography on al-Khawārizmī's algebra. The latter acts as an empty language in which both arithmetic and geometry can be expressed. The notion of "*qiyās*" (analogy) was originally a formal rule of grammar, then of jurisprudence, before it became a formal rule of algebra. This formalisation opened the way, with the successors of al-Khawārizmī, to research in algebraic proofs.

In her two writings, Radelet-de Grave highlights the fertility of interdisciplinary openness from a wider diachronic perspective. In "Art and Mathematics, Two Different Paths to the same Truth", Radelet-de Grave analyses the classifications of Arabic abstract designs made by Hermann Weyl (1885–1955) and Andreas Speiser (1885–1970) based on the symmetries that organise them. For example, it is about abstract designs of Alhambra fortress in Muslim Spain (thirteenth-fourteenth centuries), which belong to a tradition of geometric motifs that goes back to ancient Egypt. The work of classification of groups made by Weyl and Speiser contributed largely to the spreading out of group theory in twentieth century mathematics.

The notions of group and of symmetry are deeply connected. A symmetry group is the set of all geometrical transformations under which the group remains unchanged or invariant. The fundamental mathematical idea of Arab-Islamic designs is indeed "invariance", which means that motifs remain the same after a transformation in the plane: displacement, rotation or reflection. Radelet-de Grave argues that they were the product of a deep mathematical reflection, observing that all possible transformations of certain symmetry groups can be found in Arab geometric designs.

In her article "The prehistory of the Principle of Relativity", Radelet-de Grave shows that the same notion of "invariance" was essential to the emergence of a crucial notion of modern physics, namely Galilean relativity which says that the fundamental laws of physics are invariant in all frames of reference moving with constant velocity with respect to each other. Galileo (1564–1642) demonstrated it by carrying out various experiments on a ship; first when the ship was at rest, and then he repeated the same experiments when the ship was moving with constant velocity and he observed that the results of the experiments are the same in both cases. The prehistory of the notion of invariance, essential to this principle can be found in Arab designs, although their invariance does not refer to transformations in the physical space but in the geometric plane. The notion of invariance has been applied subsequently by Huygens (1629–1695) to obtain the laws of collisions, by Lorenz (1853–1928) to obtain the coordinate transformations for space-time and by Einstein (1879–1955) to formulate special relativity then general relativity.

Relying upon her analysis of the origin of invariance in Arab geometric designs, Radelet-de Grave advances a general hypothesis of philosophy of science, according to which fundamental scientific ideas are perennial and universal, although, dare we say, they can be formally theorised only when they find mature and favorable historical conditions.

Our article on "Intersections between Social and Scientific Thought. The Notion of *Muṭābaqa* in the *Muqaddima* of Ibn Khaldūn" points out that the materialist historiography of Ibn Khaldūn (1332–1406) did not always distinguish clearly between the social laws, the mathematical laws and the physical laws. In an analogous manner, the European materialist philosophy during the seventeenth and the eighteenth centuries (Hobbes, Locke, Hume) did not clearly distinguish between the laws of nature on one side and the laws of man and society on the other side. In Europe, this feature characterised an age of transition toward a modern division of labour between the natural sciences, which use the experimental scientific method, and the social sciences, which deal with historical facts that change qualitatively and quantitatively in time. If in Europe materialistic philosophies could engage in this transition with success thanks to the favourable circumstances of emerging capitalism, this was not the case for the Arab society at the time of Ibn Khaldūn, surrounded by social decay and human distrust.

Finally, the article by Paul Buell on "Arabic Medicine in China: Context and Content" deals with the reception of Arab-Greek medicine, a physical science within the Aristotelian pattern, by the Yuan Mongol Muslim dynasty in China (thirteenth century) and beyond. The article intends to highlight the positive role of the Mongols in spreading goods, technologies and ideas, and reminds us that classical Islam was engaged in international exchanges on a world scale.

To conclude, this volume presents various aspects of the history of Arabic mathematics and physics in the classical era from a comparative perspective. It expounds on advances and ruptures of classical Islam with respect to Hellenism, and namely the emergence of a new mathematical rationality, which was both algebraic and experimental, and the beginning of the mathematisation of physics. These were the features of classical science as it was understood and practiced in Europe until the end of the seventeenth century, after Europe had engaged in its own appropriation of Arabic and Hellenistic mathematics and physics as early as the twelfth century. Advances and ruptures varied from one discipline to another and from one branch to another within the same discipline. Also, we have pointed out that the interdisciplinary openness of classical Islam was a factor of creativity and contributed to advances in the scientific and social thought.

Classical Islam, which extended from Muslim Spain to the borders of China, was centered in the Near East, where it replaced and developed the Hellenistic civilisation that had flourished after the unification of that region by Alexander the Great. It was in the ninth century-Baghdad that translation of the Hellenistic heritage into Arabic took place. Classical Islam, like Hellenism, benefited from a unique geographical position for being the only region directly connected with all the areas of the ancient world: China, India, Western Europe and Sub-Saharan Africa. The transmission and development of scientific thought was centered in the Near East and gradually moved to the Western side of the Mediterranean as result of Europe's appropriation of the Arabic and Hellenistic heritage. However, technologies, goods and ideas travelled on a world-scale, and namely through the maritime and terrestrial trade routes connecting Eastern and Southern Asia with the Near East. Buell's article in this volume reminds us of the importance of these routes in the transmission of Greek-Arabic medicine to China.

From the fifteenth century onward, as a consequence of the circumnavigation of Africa, the opening of new maritime routes enabled Western Europe to be connected with Eastern and Southern Asia without passing by the intermediary of the Near East. The latter lost progressively the centuries-long benefit of its unique geographical position. In the second half of the eighteenth century, Western Europe was transformed by an unprecedented political, economic, social and cultural development. This was the framework in which, thanks to Newton's unification of mechanics, magnetism and optics, modern science came to be. For its unifying project, the potential of its technological applications, the support it received from scientific institutions and its impact on the philosophical thought, modern science, and particularly mathematics and physics, differed from classical science. After having been European for three centuries, modern science was gradually appropriated by the rest of the world, until it became what it is today: global science.

We would like to express our profound gratitude to Roshdi Rashed. Although the content of this volume does not necessarily engage him (except for his contributions), without his support and advices this work could not have been completed.

Giovanna Lelli

CHAPTER 1

Science in Islam and Classical Modernity

Roshdi Rashed

1 When Was the Renaissance?

In 1936, the German philosopher Edmund Husserl (1859–1938) wrote in his familiar style: "It is well known that during the Renaissance, European humanity underwent a revolutionary turnaround: against the prevailing Middle Ages modes of existence which it now no longer valued, preferring instead a new kind of freedom."[1] By "Renaissance" Husserl does not so much refer to the concept used by fifteenth century Italian literary and humanist circles, nor the concept found later in the writings of Erasmus (1466–1536) (where it is related to the renewal of education and religion) as to a concept related to science and philosophy, both closely linked, a concept which would mean more at the end of the sixteenth and during the seventeenth century.

The concept in question therefore appears to be associated with classical science, (i.e. early modern science), and makes two claims: it is a weapon of war and a means of explanation, or at least of description. As a weapon of war, it was used by both seventeenth century scientists and philosophers in order to mark a safe distance, real or imaginary, from the Ancients, and in order to promote their own contribution: one only has to think of Bacon (1561–1626), Descartes (1596–1650) or Galileo (1596–1650). As a means of description or explanation, the term "Renaissance"—as made perfectly clear by Husserl—does not explain or describe a completely conventional period here, but just one moment in the intellectual liberation of Europe as it tore itself away from ignorance and superstition.

But Husserl's statement is not at odds with his time: other philosophers and historians believed just as firmly that "Renaissance", "Reform" and "Scientific Revolution" were the most appropriate conceptual terms to describe classical modernity. Almost universally adopted, this view had roots that went back to the eighteenth century where it was first used to introduce the concept of "indefinite progress" as with William Wotton (1666–1727) in England and Bern-

1 Husserl, *La crise des sciences européennes et la phénoménologie transcendantale* (1976), 12.

 | DOI:10.1163/9789004513402_003

ard de Fontenelle (1657–1757) in France. In the nineteenth century, German Romanticism gave it an anthropological dimension which it had not previously possessed. But regardless of its roots, this belief poses a central question as to the origins and development of classical modernity, which is closely related to science and its philosophy.

Behind the apparent unanimity, this belief had already come under the attack of a fellow scholar of the German philological school, Pierre Duhem (1861–1916). The fact is, his philosophy of science, as well as his religious and political beliefs meant that he, a famous French physicist and historian of Latin medieval science, had a far better grasp of historical continuity and the appeal of the Middle Ages. He therefore dates this classical modernity as far back as the fourteenth century with the Latins, i.e. Merton College, Oxford, and Paris University. This thesis has since been contested by historians of ideas and of science such as Charles H. Haskins (1870–1937), Alexandre Koyré (1882–1964), George Sarton (1884–1956), etc. It has also been contested, but in a different way, in the exceptional works of Annelise Maier (1905–1971). More recently, Marshall Clagett (1916–2005) has attempted to balance the argument.

Yet this debate, and the efforts of many scholars during the course of this century, have made it clear that concepts such as "Renaissance", "Reform" and "Scientific Revolution" cannot account for the accumulated facts, and that in the evolution of classical science, the fourteenth century has been somewhat eclipsed by the twelfth and thirteenth centuries, when the Latins started to make Hellenistic science and Arabic science their own—and this is in fact three centuries before the "Renaissance". Traditional methods of dividing political or cultural periods therefore prove inadequate when it is a matter of understanding and analysing classical modernity. Original Islamic works of science are themselves not included here, but they are referred to in their Latin translation, and in this way, maintain a presence in the debate.

2 Science in Islam

It is this last matter that I should like to take up now—science in Islam (no longer limited to Latin translation alone) and classical science. My aim is to examine how knowledge of Arabic science can bring about a better understanding of classical science both epistemologically and historically. Two features that characterise classical science will be considered: (a) new mathematical rationality and (b) experiments as a category of proof.

2.1 *The New Mathematical Rationality*

I now turn, not to a philosopher such as Husserl, but a simple barber, the barber of Baghdad, who expresses himself thus in *The Arabian Nights*:

> … and you find in me the best barber in Baghdad, an experienced doctor, a profound chemist, an astrologer who does not make mistakes, an accomplished grammarian, a perfect rhetorician, a subtle logician, a mathematician accomplished in geometry, in arithmetic, in astronomy and in all the refinements of algebra; a historian who knows the history of all the kingdoms of the Universe. Beyond this, I am in command of all the parts of philosophy; I have in my memory all our laws and all our traditions, I am a poet, an architect …[2]

One can see how both mathematics and algebra—in its own right and with all its refinements—occupied a prime position in the Encyclopaedia of popular knowledge in the great cities of the time: the Barber echoes classifications of far more learned sciences; classifications of, amongst many others, the tenth-century philosopher al-Fārābī, or, in the next century Ibn Sīnā, which, unlike other Greek or Hellenistic classifications, welcomed a new independent discipline and gave it its own title: algebra. The popularity of mathematics, its spread and the privileged role of algebra are features of what may be called Arabic science.

Let us briefly go through the genesis of the main features of these Arabic mathematics and, to do so, let us go back to Baghdad in the early part of the ninth century. The process of translation of the great Hellenistic mathematical compositions was at its height and presented two striking characteristics:

(1) Translations are the work of mathematicians, often top ones, and are prompted by the most advanced research of the time.
(2) This research was not prompted by theoretical interests alone, but by the needs of a new society in the fields of astronomy, optics, arithmetic, its need for new measuring instruments, etc.

The early ninth century is therefore a great moment of expansion of Hellenistic mathematics in Arabic. And it is at precisely that time and from within the elite circle of the "House of Wisdom" in Baghdad that Muḥammad ibn Mūsā al-Khawārizmī (ca. 780–850) writes a book on a subject and in a style which are both new. It is in those pages that algebra features for the first time as a distinct and independent mathematical discipline. The event was crucial and

2 *Le Milles et Une Nuits* (1965), I, 426–427.

perceived as such by al-Khawārizmī's contemporaries, as much for the style of mathematics as for the ontology of the subject, and, even more, for the wealth of possibilities that it offered from then on. The style is both algorithmic and demonstrative, and already we have here, with this algebra, an indication of the immense potential which will pervade mathematics from the ninth century onwards: the application of mathematical disciplines one to another. In other words, algebra, because of its style and generality of purpose, made these interdisciplinary applications possible, and they in turn, by virtue of their number and diversity, will, after the ninth century, constantly modify the structure of mathematics. A new mathematical rationality is born, one that we think will come to characterise classical mathematics, and more generally, classical science.

Al-Khawārizmī's successors began—bit by bit—to apply arithmetic to algebra, and algebra to arithmetic, and both to trigonometry; algebra to Euclid's theory of numbers, algebra to geometry and geometry to algebra. These applications were the founding acts of new disciplines, or at least of new chapters. This is how polynomial algebra came to be; as well as combinatorial analysis, numerical analysis, the numerical resolution of equations, the new theory of numbers and the geometric construction of equations. There were other effects as a result of these multiple applications—such as the separation of integer Diophantine analysis from rational Diophantine analysis, which would eventually have a chapter of its own within algebra under the title of "indeterminate analysis".

From the ninth century onwards, therefore, the mathematical landscape is never quite the same: it is transformed, its horizons widen. One first sees the extension of Hellenistic arithmetic and geometry: the theory of conics, the theory of parallels, projective studies, Archimedean methods of measuring surfaces and curved volumes, isoperimetrical problems, geometrical transformations; all these areas become subjects of study for the most prestigious of mathematicians (Thābit ibn Qurra, Ibn Sahl, Ibn al-Haytham, to name but a few) who manage, after in-depth research, to develop them in the same fashion as their predecessors, or by modifying them whenever necessary. At the same time, within the tradition of Hellenistic mathematics, there is seen to be an exploration of non-Hellenistic mathematical areas.

It is this new landscape, with its language, its techniques and its norms, which will gradually become the landscape of the Mediterranean. Let us take two examples: rational Diophantine analysis and integer Diophantine analysis.

2.1.1 Rational Diophantine Analysis

The emergence of indeterminate analysis—or, as it is called today, Diophantine analysis—as a distinct chapter in the history of algebra, goes back to the successors of al-Khawārizmī, and namely to Abū Kāmil. His book, written around 880, was translated into Latin in the twelfth century and into Hebrew in the fifteenth century in Italy.

Abū Kāmil (ninth-tenth century)'s purpose in his *Algebra* is to improve upon previous uncoordinated works, and to give a more systematic account; including not only problems and their algorithm solutions, but methods as well. Indeed, Abū Kāmil, towards the end of his *Algebra*, deals with 38 Diophantine problems of the second degree and the systems of these equations, 4 systems of indeterminate linear equations, other systems of determinate linear equations, a group of problems centred around arithmetical progression, and a further study of this last group.[3] This collection satisfies the double goal set by Abū Kāmil: to solve indeterminate problems and at the same time to use algebra to solve problems that arithmeticians usually dealt with. In Abū Kāmil's *Algebra*, for the first time in history as far as I know, there is an explicit distinction drawn between determinate and indeterminate problems. A study of his 38 Diophantine problems not only reflects this distinction, it also shows that the problems do not succeed each other randomly, but according to an order implicitly indicated by Abū Kāmil. He puts the first 25 all into the same group, and gives a necessary and sufficient condition to determine rational positive solutions. Thus for instance

$$x^2 + 5 = y^2$$

Abū Kāmil reduces the problem to that of dividing a number the sum of two squares into two other squares and solves it. Abū Kāmil's techniques of resolution show that he knows that if one of the variables can be expressed as a rational function of the other, or, more generally, if a rational parameterage is possible, then all solutions are possible. Whereas if, on the other hand, the sum has led to an expression with an unresolvable radical, then there is absolutely no solution. In other words, unknown to Abū Kāmil, a second degree curve does not possess a rational point, nor is it bi-rationally equivalent to a straight line.

---

3 Abū Kāmil, *Kitāb fī al-jabr wa al-muqābala* [*The Book of Algebra*] (MS 379, Istanbul: Kara Mustafa Paşa Collection, Beyazit Library).

The second group is made up of 13 problems that are impossible to parameterise rationally. Once more, in a language unknown to Abū Kāmil, they all define curves of genus 1, as for instance the problem:

$$x^2 + x = y^2,$$
$$x^2 + 1 = z^2,$$

which defines a "skew quartic" curve of $\mathbf{A}^3$ of genus 1.

Half a century later, al-Karajī (980–1030), another algebraist, extends rational Diophantine analysis further than ever before. He marks an important point in the history of algebra by formulating the concept of polynome and algebraic calculus of polynomials. In rational Diophantine analysis, al-Karajī differs from his predecessors—from Diophantus to Abū Kāmil—in that he does not give well-ordered lists of problems and their solutions, but instead structures his account on the basis of the number of terms in the algebraic expression, and on the difference between their powers. Al-Karajī considers, for example, successively

$$ax^{2n} \pm bx^{2n-1} = y^2, \;\; ax^{2n} + bx^{2n-2} = y^2, \;\; ax^2 + bx + c = y^2$$

This is a principle of organisation which would be borrowed by his successors. Al-Karajī further advances the task initially undertaken by Abū Kāmil, highlighting—as far as is possible—the methods for each class of problems. We can show the problem which defines a curve of genus 1 in $\mathbf{A}^3$ simply as:

$$x^2 + a = y^2$$
$$x^2 - b = z^2$$

Al-Karajī's successors have attempted to follow the path that he laid out. I shall not elaborate further on the matter of rational Diophantine analysis in Arabic, and will return to the development of integer Diophantine analysis.

2.1.2 Integer Diophantine Analysis

The tenth century sees for the first time the constitution of integer Diophantine analysis, or new Diophantine analysis, doubtless thanks to algebra, but also, in some ways, despite it. The study of Diophantine problems had been approached on the one hand by demanding integer solutions, and on the other by proceeding according to demonstrations of the type found in Euclid's arithmetical books of Elements. It is the specific combination—for the first time in history—of the realm of positive integers (understood as line segments), algebraic techniques and pure Euclidean-style demonstration that permitted the birth of the new Diophantine analysis. The translation of Diophantus's *Arithmetica*, as we know, provided these mathematicians not so much with methods as with problems in the theory of numbers which they found formulated therein. Unlike their Alexandrine predecessor, they wasted no time in systematising and examining these problems: the representation of a number which is the sum of squares, congruent numbers, etc.

This is how tenth century mathematicians such as al-Khāzin studied numerical rectangular triangles and problems of congruent numbers. Al-Khāzin gives the theorem of congruent numbers as follows:[4]

Given a natural integer a *the following conditions are equivalent:*

1° the system

$$x^2 + a = y^2$$
$$x^2 - a = z^2$$

admits a solution;

2° there exist a couple of integers (m, n) such as

$$m^2 + n^2 = x^2,$$
$$2mn = a;$$

in these conditions, a is in the form $4uv(u^2 - v^2)$.

It was also in this tradition that the study of the representation of an integer as the sum of squares started: in fact, al-Khāzin devotes several propositions in his dissertation to this study. These tenth century mathematicians were the first to address the question of impossible problems, such as the first case of Fermat's theorem. But in spite of all their efforts, this problem continued to occupy mathematicians, who later stated the impossibility of the second case, $x^4 + y^4 = z^4$.

Research into integer Diophantine analysis did not die with its initiators after the first half of the tenth century: quite the contrary, their successors carried on, at first in the same spirit. But, towards the end of its evolution, there was a noticeable increase in the use of purely arithmetical means in the study of Diophantine equations.[5]

4 Rashed (1984).
5 See Rashed, "Al-Yazdī et l'équation $\sum_{i=1}^{n} x_i^2 = x^2$" (1994), 79–101.

2.1.3 The Tradition Continues

With this example of Diophantine analysis, I wished to illustrate how algebra conceived at the time of al-Khawārizmī was central to the foundation and transformation of this new discipline. As we have seen, the dialectic between algebra and arithmetic has meant that rational Diophantine analysis was considered as part of algebra. And from then on, from al-Karajī to Euler (1707–1783), an important treatise of algebra would always include a chapter on rational Diophantine analysis. This stage marks the birth of integer Diophantine analysis, which would be bound to comply with the exigencies of demonstration. With these disciplines, we have finally seen the rise of elements of a new mathematical rationality which admits the infinity of solutions as a genuine solution. This allows us to differentiate between several types of infinity of solutions—such as the identities and infinitely great numbers—and to positively consider impossibility, or impossible solutions as a subject for construction and demonstration.[6] However, all these features are precisely those of classical Diophantine analysis as it was conceived and practised in the seventeenth century by Bachet de Méziriac (1581–1638) and Fermat (1601–1665). Around 1640, Fermat invents the method of infinite descent[7] which itself would breathe new life into the discipline, but that is another story.

One might ask whether this so-called epistemological continuity corresponds to a particular historical continuity and, if so, to which? To put it more bluntly, was Bachet de Méziriac, at the beginning of the seventeenth century, created out of nothing? Let us ponder this question for a while, as it affects our subject. My answer would be to simply recall the figure of one of the most prominent Latin mathematicians of the Middle-Ages and the source of many Renaissance writings: Fibonacci, alias Leonardo Pisano (1170–after 1240).

Fibonacci who lived in Bougie and who travelled in Syria, Egypt and Sicily, was in touch with Emperor Frederic II (r. 1215–1250) and his court. This court included Arabists dealing with Arabic mathematics, like John of Palermo, and Arabic speakers knowledgeable in mathematics, like Théodore of Antioch. Fibonacci wrote a Diophantine analysis, the *Liber Quadratorum*, that historians of mathematics rightly hold to be the most important contribution to Latin Middle-Ages theory of numbers, before Bachet de Méziriac and Fermat's contributions. The purpose of this book, as stated by Fibonacci himself, is to solve this system:

6 Rashed (1984), 195 f.

7 Itard, *Essais d'histoire des mathématiques* (1984), 229–234.

$$x^2 + 5 = y^2$$
$$x^2 - 5 = z^2$$

proposed by John of Palermo. This is not just any question of Diophantine analysis, but a problem that crops up as a problem in its own right in the works of al-Karajī and many others. More generally, the main results revealed in the *Liber Quadratorum* are either those obtained by Arabic mathematicians in the tenth and eleventh centuries, or are very close to those. Furthermore, the results are placed in an identical mathematical context, namely the theory of Pythagorean triplets, so the conclusion is really nothing new; a prominent historian whose admiration for Fibonacci cannot be doubted had already put it forward. I am referring to Gino Loria who wrote:

> It seems difficult to deny that Leonardo of Pisa (Pisano) has been led to research that had already been summarised by Muhammad ibn Ḥosein (read al-Khāzin), and his dependence on him is even more in evidence in the following section of the *Liber Quadratorum* which deals with "congruent numbers".[8]

We can see therefore that the *Liber Quadratorum* truly belongs to the tradition of tenth century mathematicians, who created integer Diophantine analysis.

Although the case of Fibonacci and Diophantine analysis is not unique, it is exemplary, considering the level it reached. This mathematician, looked at from one direction can be seen as one of the great figures in Arabic mathematics of the ninth to eleventh centuries, but, looked at from another direction, can be seen as a scholar of fifteenth to seventeenth century Latin mathematics.

We have seen in this example how classical scientific modernity had its roots in the ninth century, and that it continued to develop until the late seventeenth century. In this way, rational Diophantine analysis lives on into the eighteenth century, whereas integer Diophantine analysis undergoes a new revolution in the mid- seventeenth century. We also see that this modernity is written about in Arabic in the early stages, that it was then transmitted through Latin, Hebrew and Italian, before going on to become part of significant new research. And finally, we see that the rational core of this modernity was algebra, and that the conditions which allowed it to exist are inherent in the new ontology contained within its discipline.

8 Loria, *Storia delle matematiche dall'alba della civiltà al secolo XIX* [A History of Mathematics from the Dawn of Civilisation to the Nineteenth Century] (1950), 234.

With this description, we are very far from the prevailing global attitude, and the term "Renaissance" seems a rather an inadequate one to describe the mathematical facts.

2.2 *Experiments as a Category of Proof*

Let us now look at the second feature of classical scientific modernity. I refer here to experimental norms such as norms of proof. To put it briefly, the reduction of the separation between science and art, as well as the change in their relationship in Islamic civilisation—a civilisation which was far more urbanised than the previous civilisations—resulted largely in the extension of empirical research, and the genesis of a vague idea of experimentation. Thereafter, the systematic use of empirical procedures increased; for example, in botanical and linguistic classifications, physicians' control experiments and alchemists' experiments, physicians' clinical observations and comparative diagnostics.

But it is necessary to wait for the establishment of new relations between mathematics and physics before we see the still-unclear notion of experimentation granted the dimension that would define it: systematic and ordered proof. This concept was completely new, and should not be confused with that of controlled, or even measured observation, in astronomy. This time, it will become necessary to take into account the very existential nature of examined phenomena. Optics was the first discipline where such an idea first saw the light of day, before being further worked on in mechanics. The concept emerged for the first time in this form in the works of Ibn al-Haytham (965–1040), specifically, in his book *Optics*, translated into Latin in the twelfth century, and later into Italian. Republished by Friedrich Risner in the sixteenth century, it was the reference book of all scholars during the Middle-Ages, and later for Kepler (1571–1630), Descartes (1596–1650) and Malebranche (1638–1715), among many others.[9]

In order to understand the emergence of the new norms and practices, we should briefly recall Ibn al-Haytham's project. He was involved for the greater part of his writings in effecting a programme of reform within the discipline, which was precisely what led him to revisit the various different disciplines, one by one: optics, meteorological optics, catoptrics, burning mirrors, dioptrics, burning sphere, physical optics. The fundamental act of this reform consisted in clearly distinguishing between conditions for propagation of light and conditions for viewing objects. This reform has on the one hand lent physical sup-

9 On the optic works of Ibn al-Haytham, see Rashed (1992) and Rashed (1993).

port to the rules of propagation (in the case of a true mathematical analogy between a mechanical model of the movement of a solid ball thrown against an obstacle, and a similar movement of light) and on the other hand led to the use of geometrical procedures and experimentation everywhere.

Optics no longer meant what it had meant to the Greeks: that is, a geometry of perception. It now comes in two parts: a theory of vision, to which the physiology of the eye and the psychology of perception are also linked; and a theory of light, to which geometrical optics and physical optics are linked. This reform has led to—among other things—the emergence of new questions, never before formulated. For example, examination of the spherical lens and the spherical diopter, not only as burning devices, but also as optical instruments, in dioptrics. It also led to the creation of a new practice of investigation, together with a new lexicon, that of experimentation.

But what does Ibn al-Haytham mean by "experimentation"? One finds in the works of Ibn al-Haytham as many different meanings of the word, and as many different functions assumed by experimentation as there are relations between mathematics and physics. These relations are effectively established according to various modes; they are not organised around a particular theme by Ibn al-Haytham, but they are implicit in the various works, and it is possible to analyse them.

The reform of geometric optics is Ibn al-Haytham's main contribution to the field: and the unique relationship between mathematics and physics in geometric optics is an isomorphism of structures. But because he had already defined a beam of light, Ibn al-Haytham was able to write about the phenomena of propagation, including the phenomenon of diffusion, in such a way that they perfectly fit the laws of geometry. Several experimental devices were invented as a technical check for propositions whose language had already been controlled by geometry: experiments designed to test the laws and rules of geometrical optics. A reading of the works of Ibn al-Haytham testifies above all to two important facts. First of all, Ibn al-Haytham's experiments were not designed merely to test qualitative assertions, but also to obtain quantitative results. Secondly, the devices conceived by Ibn al-Haytham, were varied and complex for their time, and were not limited to those used by astronomers.

In physical optics, there is another type of relationship to be found between mathematics and physics, and later a second meaning of the word "experimentation". The contribution of mathematics at this stage is in the analogies between diagrams of the movement of a heavy body and diagrams of reflection and refraction. In other words, mathematics was introduced into physical optics by means of dynamic diagrams of the movement of heavy bodies that were presumed to be already mathematised. It is precisely this preliminary mathematisation of the concepts of physical doctrine that allowed them to be transferred to the level of an experimental situation. Although this was definitely a temporary stage, it nevertheless provided a level of existence for concepts which—although semantically indeterminate—were syntactically structured, such as Ibn al-Haytham's diagram of projectile movement, later taken up by Kepler and Descartes.

A third type of experimentation, not practised by Ibn al-Haytham himself, but made possible by his reforms and discoveries in optics, appears at the end of the thirteenth century in the works of his successor al-Fārisī. In this case, the established relationship between mathematics and physics tends towards the construction of a model, then subsequently, through geometry, towards the systematic reduction of the propagation of light in a natural environment to its propagation in a manufactured object: it becomes a matter of defining a truly mathematical analogical correspondence between the natural and the manufactured object, as for example, the model of a massive glass sphere filled with water, to explain the rainbow. Experimentation here has the function of expressing the physical conditions of a phenomenon that could not otherwise be studied either directly or completely.

Other examples could be added to these three types of experimentation, but it is enough to say that—despite the difference in functions they fulfil—the three types of experimentation that we have just studied are all control mechanisms as well as levels of existence for syntactically structured notions. In this they are significantly different from observation—even traditional astronomical observation. In all three types we are in situations where the scientist intends to physically construct the subject himself, in order to be able to think about it: physically realising an idea that could not previously be realised.

Ibn al-Haytham's reforms survived him, as did his establishing of experimental norms as an integral part of a physics proof. From Ibn al-Haytham down through Kepler and other seventeenth century scientists, the line of descent is established. And a knowledge of Arabic science is necessary for our understanding of classical modernity: it enables us to grasp the introduction of experimental norms, and also to better understand the emergence, in the late seventeenth century, of a still-unknown dimension of experimentation, namely the quest for precision.

3 Conclusion

To conclude then, let us recall the two central points of this paper. We began by seeing that the new possibilities afforded by algebra were the origin of a new strategy and of a new rationality. This strategy was inherent in algebra's own development after al-Khawārizmī, as well as in its relationship with other mathematical disciplines. It is a strategy that consisted of increasingly exhibiting structures and operations in algebra, and initiating the previously-mentioned dialectic of application in its relationship with other disciplines. As far as the new rationality is concerned, it is based on a new ontology of mathematical subjects, making possible what was not possible before. For example, the same subject could be determined both geometrically and arithmetically; a problem could have an infinite number of true solutions; an approximate solution could be a true solution; an impossible solution could also be a true solution; the same procedure could be applied to different objects without additional justification, etc.

We have also witnessed the rise of the new concept of proof in physics, and we have seen how, from then on, it was accepted that the level on which a physical object existed was no longer its "natural" level, but was within the realms of the experimental.

This new rationality, which can in summary be called algebraic and experimental, characterises classical modernity, and was, as we have said, founded between the nineth and twelfth centuries by scholars as far apart as Muslim Spain and China, all of whom were writing in Arabic. Appropriation of this new rationality by scholars began in the twelfth century, and a new improved version was to appear from the sixteenth century onwards. It would therefore seem that whoever wishes to understand classical modernity should not subscribe to the historian's idea of periods or eras, since these are founded on causal links between events of political, religious and literary Renaissance history and events in science: rather, he or she should go in search of true paths, and leave aside the myths and legends which led such great minds as Husserl's astray.

Bibliography

Abū Kāmil (ninth-tenth century), *Kitāb fī al-jabr wa al-muqābala* [*The Book of Algebra*] (MS 379, Istanbul: Kara Mustafa Paşa Collection, Beyazit Library, fol. 79r–110v.).

Husserl, Edmund, *La crise des sciences européennes et la phénoménologie transcendantale*, transl. Gérard Granel & Jacques Derrida (Paris: Gallimard, 1976).

Itard, Jean, *Essais d'histoire des mathématiques*, collected and introduced by Roshdi Rashed (Paris: Librairie scientifique et technique A. Blanchard, 1984).

Le Milles et Une Nuit, transl. Antoine Galland, Vols. I–III (Paris: Garnier Flammarion, 1965).

Loria, Gino, *Storia delle matematiche dall'alba della civilità al secolo xix* [A History of Mathematics from the Dawn of Civilisation to the Nineteenth Century] (Milano: Hoepli, 1950).

Rashed, Roshdi, *Mathématiques et optique: recherches sur la pensée scientifique en arabe* (Aldershot: Variorum, 1992).

Rashed, Roshdi, *Géométrie et Dioptrique au xe siècle. Ibn Sahl, al-Qūhī et Ibn al-Haytham* (Paris: Les Belles-Lettres, 1993).

Rashed, Roshdi, "Al-Yazdī et l'équation $\sum_{i=1}^{n} x_i^2 = x^2$", *Historia Scientiarum*, 4: 2 (1994), 79–101.

Rashed, Roshdi, *Entre arithmétique et algèbre: Recherches sur l'histoire des mathématiques arabes* (Paris: Les Belles Lettres, 1984), English translation *The Development of Arabic Mathematics: Between Arithmetic and Algebra*, Studies in Philosophy of Science, (Boston: Kluwer, 1994).

CHAPTER 2

Physics and the Mathematical Sciences in the Islamic Period: a Conceptual and Historical Survey

Hossein Masoumi Hamedani

The aim of this article is to trace the history of the conceptual and historical relations between two groups of scientific disciplines, the mathematical sciences and physics, during the Islamic period. In so doing, I hope to show how sophisticated these relations were and how the acknowledgement of the existence of some disciplines with both a mathematical and a physical aspect created a deep cleavage within the Aristotelian conception of science. I will discuss the different reactions of mathematicians and physicists of the Islamic period to this situation and, at the end, I will investigate the Alhazenian idea of the necessity of the "composition" of the physical and mathematical approaches to some problems.

1 Physics and the Mathematical Sciences: the Conceptual Problem

Of the two constituent parts of the term "mathematical sciences", one is relatively new and the other very old. The use of the term "science" to designate a privileged type of knowledge characterised by its method is relatively recent, whereas the use of the term "mathematics" to refer to a group of disciplines is as old as philosophy itself.

In the works of Aristotle (384–322 BCE), we find many references to mathematics and the mathematical sciences. In fact, the latter are an integral part of his classification of philosophical disciplines, occupying a middle place between physics[1] and metaphysics. At the time of Aristotle, this group of sciences, i.e. the mathematical sciences, comprehended not only arithmetic and geometry, but

1 The term "physics", as used by Aristotle and his followers, had a range of meaning much wider than what we mean by this term today. It not only covered the general laws governing the motion of all inanimate bodies, including heavenly bodies, but also all biological sciences and the science of the soul. Moreover, as we will see in this article, its method was totally different from the method of modern, i.e. Newtonian and post-Newtonian, physics.

 | DOI:10.1163/9789004513402_004

also optics, harmonics and astronomy.[2] According to Aristotle, these branches of mathematics, i.e. optics, harmonics and astronomy, "come nearest to the study of nature", because:

> They are in a way the reverse of geometry, geometry considers natural lines, but not as natural; optics treats of mathematical lines, but considers them not as mathematical but as natural.[3]

Regarding this text of Aristotle, two remarks seem to be in order. In the first place, it would seem that this Aristotelian thesis is somehow inconsistent with the criteria he uses in his classification of sciences. In this classification the three main divisions of science are ordered according to the manner in which their subject-matter can be separated from matter. The subject-matter of metaphysics (which Aristotle calls theology or the first philosophy) is separable from matter both in reality and in mind. The subject-matter of mathematics is separable from matter only in mind; in external reality it exists always in matter. The subject-matter of physics cannot exist in separation from matter, neither in mind nor in external reality. Now in the passage cited above, Aristotle says that optics, as a mathematical science, considers its subject-matter, that is, the visual rays, "not as mathematical but as natural." Does he intend to say that optics is not a mathematical but a physical science? If so, why does he talk of opticians as a group of scholars distinct from natural philosophers (physicists)?

The second point is the epistemic status of these disciplines, that is, the kind of knowledge they can provide, its degree of certainty and its relation to the knowledge provided by physics. This point is crucial, because certain mathematical sciences, in the form they had at the time of Aristotle, were based on premises which Aristotle himself thought to be false. The most notorious example is the science of optics, mentioned by Aristotle in the text quoted above. At the time of Aristotle, this "mathematical" discipline, which aimed to account for the phenomenon of vision by the use of geometry, was based on the assumption that vision takes place through the intermediary of visual rays emanating from the eye, each ray being represented by a straight line. On several occasions, Aristotle invokes this science, as well as the thesis of the existence of visual rays on which it was based, to account for some phenomena, such as the halo and the rainbow.[4] Nevertheless, in Aristotle's own view, vision

---

2 Aristotle, *Physics* 194a, II, (1970), 2, 7–12.

3 Aristotle, *Physics* 194a, II (1970), 2, 7–12.

4 Aristotle, *Meteorology* 373a–375b, in *The Complete Works of Aristotle* (1984). Elsewhere, regarding someone who always saw an image in front of him when he walked, Aristotle says that he

was the result not of something material emanating from the eye (or from the object) but of the activation of the transparent medium and the transmission of the forms of the things seen to the eye. How could Aristotle base his explanation of a natural phenomenon, such as the halo or the rainbow, on an evidently false premise?

The very idea of a science based on a false assumption went against the Aristotelian conception of a demonstrative science. This conception is presented in the beginning passages of Aristotle's *Posterior Analytics* as a deductive structure based on certain premises which had to satisfy some minimal conditions. Following the French philosopher Gilles Gaston Granger, we can divide these conditions into two classes. The first class, which Granger calls "absolute conditions", concerns each premise of a science taken in isolation. Seen under this angle, the premises of a science should satisfy three conditions. They have to be 1) true, 2) immediate, and 3) primitive. The second class of conditions, which Granger calls "relative conditions", concerns the characteristics of the premises in their relation to the conclusions drawn from them. From this point of view, the premises should be 4) more familiar than the conclusion, 5) prior to the conclusion, and finally 6) the cause of the conclusion.[5]

Let us put aside for the time being the other conditions and examine only the first and the last. The first condition, i.e., that the premises of a science have to be true, seems intuitively valid. Few are prepared to accept as scientific a false statement or a statement obtained from false premises through logical deduction. The sixth condition, that the premises should be the cause of the conclusion, underlines one of the main characteristics of every scientific explanation in the Aristotelian conception of science. According to Aristotle, to know is to know the cause, so every veritable explanation should be a causal one. As we will see later, it was exactly the absence of this causal relation which distinguished a mathematical from a physical explanation.

Let us return for the moment to the first condition. The fact that, according to Aristotle, the first condition to be satisfied by the premises of a science is truth, and the fact that despite this condition Aristotle himself admits some results obtained by the science of optics of his time, which was, as we said before, based on a false premise, has created a great problem for historians and philosophers of science. Much has been written about this problem. A tradition of

saw this image because the visual rays emanating from his eye were so weak that they could not travel too far and they reflected back to his eye, so that the air opposite the person acted as a mirror which reflected his visual rays towards him and the image he saw was in fact his own image formed in the air (Aristotle, ibid. 373b, 1–10).

5 Granger (1976), 73–75.

interpretation which goes back at least to the nineteenth century's French historian and philosopher of science Pierre Duhem[6] and is supported by some contemporary historians of science such as the late Gerard Simon and A. Mark Smith,[7] believes that, because of their failure to meet the first prerequisite of a demonstrative science, ancient and medieval mathematical sciences (especially astronomy and optics) cannot be considered as real demonstrative sciences in the Aristotelian sense. Being sciences of appearances, they had nothing to do with reality as it is, but only with the way things appear to the eye.

In this way the advocates of this interpretation believe to have solved the problem of the relation between physics and at least some mathematical sciences: no conflict exists between these two branches of learning, because they have different subject-matters and follow different aims, one is concerned with reality as it is and the other with "saving the phenomena". I think that this line of interpretation cannot do justice to the problem, and in what follows I will argue that, as far as the Islamic period is concerned, the relation between the two disciplines was much more complicated.

2 Avicenna and the Classification of Mathematical Sciences

The conceptual aspect of this relation can best be seen through an examination of a text of Avicenna (Ibn Sīnā, 980–1037). In the eighth chapter of the first Book of the *Physics* of his *Shifāʾ* (*The Book of Healing*), entitled "On how the science of physics conducts investigations and what, if anything, it shares with other sciences",[8] Avicenna treats of the relations between physics and the other sciences. By "the other sciences" he means the mathematical sciences which existed at his time. The list he gives is not exhaustive and he proceeds rather by giving examples. The mathematical sciences he talks about are arithmetic, geometry, the science of the moving sphere, the science of weights, music, optics and astronomy. Nevertheless, what he says about this handful of sciences is general enough to be true for other mathematical sciences as well. In what follows, I try to present the outline of Avicenna's argument in a rather synthetic way without betraying the general line of his argument.

6 Pierre Duhem has applied this thesis, mostly for the history of medieval astronomy, in his voluminous *Le système du monde. Histoire des doctrines cosmologiques de Platon à Copernic* (1913–1959).

7 Cft. Simon (1989); Ptolemy, *Optics*, (1996). Both authors defend a version of this theory in the case of ancient and medieval optics.

8 Avicenna, *Al-samāʿ al-ṭabīʿī*, transl. McGinnis (2010), I, 54–60.

According to Avicenna, the relation between physics and the mathematical sciences can be examined under three criteria. The first criterion is their relative subject-matters. Developing an idea already present in Aristotle, Avicenna argues that "in a sense", arithmetic and geometry fall under physics, because they deal with numbers and magnitudes and these two are among the accidents (or accidents of accidents) of physical bodies. Nevertheless, Avicenna believes that arithmetic and geometry are not physical sciences in the proper sense of the word. The reason is that "in the construction of demonstrations [...] neither discipline needs to turn to natural matter or to take premises that refer to matter in any way."[9]

In spite of this separation, Avicenna believes that geometry is more physical than arithmetic, for the evident reason that geometrical categories—points, lines, surfaces—are applicable only to material bodies, whereas arithmetical categories, for example unity and multiplicity, are applicable to non-material entities as well.

The third mathematical science mentioned by Avicenna, i.e. the science of the moving sphere, comes closer to physics, because it treats of the motion of a sphere, and motion is specific to natural bodies.[10]

---

9 Avicenna, Ibid, 60.

10 Avicenna evidently refers to Autolycus's (fl. fourth century BCE) treatise *On the Moving Sphere*, which in the Hellenistic period was part of the "little astronomy" and later on became a part of the "intermediate books" studied between Euclid's *Elements* and Ptolemy's *Almagest*. The treatise is unique in its kind and, as far as I know. No one else has cited it as exemplifying a "science". Nevertheless, it is a good example of a work which tries to present a geometrical "model" for a physical phenomenon: the uniform circular motion of the heavenly sphere. This is done by a minimal use of physical concepts. The sole physical premise on which the treatise is based is the existence of a uniform circular motion and Autolycus does not even ascribe it to the heavenly sphere. The mathematical concepts which intervene in the treatise are "sphere", its "axis" and its "poles". So, even if the treatise does not claim to be about an independent science, its structure and content are so that Avicenna can find in it the best example for a mathematical science with a minimal physical content [for the structure of this treatise and its reception in the Islamic world, cfr. Nikfahm-Khubravan, Eshera (2019), 7–71]. In a treatise on the classification of sciences, entitled *The Divisions of the intellectual Sciences*, which exists in a unique manuscript, Avicenna gives a more general definition of the science of the moving sphere. Among the eight "divisions" of the theoretical part of the mathematical sciences, one division "contains the definition of moving solids, i.e. the properties which are attributed to their figures and intersections when they move. On this subject there are the books on the moving spheres by Archimedes (*sic*!) and others", Ibn Sinā, *Aqsām ʿulūm al-awāʾil* (MS 712, Tehran: Majlis, fol. 278). In his other treatise on the classification of sciences, entitled *Aqsām al-ḥikma* (*The Divisions of Philosophy*), Avicenna does not include this science among the divisions of mathematical sciences [Kadivar (winter 2009), 111–137].

Nevertheless, judged by a second criterion, none of these three sciences can be regarded as a physical science. This criterion is the premises on which each one of them is based. Neither arithmetic and geometry, nor the science of the moving sphere, receive their premises from physics. In this respect, they are different from a second group of mathematical sciences, which includes music, optics, astronomy and the science of weights. The specificity of this group of sciences is that they are based on two kinds of premises, physical premises and mathematical premises.[11]

Avicenna's division of mathematical science does not stop here. He goes on dividing this last group of sciences itself into two subgroups, and it is here that his third criterion comes in. This criterion is the existence of common problems.

By this criterion, the first subgroup consists of music, optics and the science of weights. Despite the fact that the disciplines belonging to this subgroup are based on both mathematical and physical *premises*, their *problems* are different from those investigated by physics.

As to the second subgroup, according to Avicenna, it has a sole member which is the science of astronomy, since it is the only science which has problems in common with physics.[12] These common problems are formulated by propositions in which both the subject and the predicate belong to mathematics and to physics, "as if this science were a mixture of physical and mathematical [sciences]". The example of such a problem given by Avicenna is the proposition "the Earth is spherical".[13] According to Avicenna both the physicist and the astronomer undertake to prove this proposition, their difference lying in the kind of premises they use in their proofs. While the physicist uses

11 Although Avicenna does not specify these physical premises, he seems to refer to premises such as the emanation of visual rays from the eye, which is the first premise of Euclid's *Optics* (the sole optical text he mentions in the above-mentioned *The Divisions of the intellectual Sciences*). The case is much clearer for astronomy: many astronomical books begin, following the *Almagest*, by the demonstration of the sphericity of the Earth and its central position in the universe, which according to Avicenna are physical premises [cfr. for example, Aḥmad ibn Kathīr al-Farghānī, *Jawāmiʿ ʿilm al-nujūm* (MS Princeton, Garrett, 967H, ff. 10a–16a);] Kūshyār ibn Labbān al-Jīlī, *al-Zīj al-jāmiʿ* [MS Istanbul, Süleymaniye Kütüphanesi, Yeni Cami, 784, ff. 319b–320b]. Later on, these two groups of mathematical and physical premises were specified in *hayʾa* (astronomy) books, many of which begin by two lists of the principles on which this science is based, one mathematical, the other physical [cfr. for example, Naṣīr al-Dīn al-Ṭūsī, *al-Risāla al-Muʿīniyya* (1387/2008), 25–47; Naṣīr al-Dīn al-Ṭūsī, *Al-Tadhkira fī ʿilm al-hayʾa*, ed. Jamil Ragep, *Naṣīr al-Dīn al-Ṭūsī's Memoir On Astronomy* (1993), I, 90–107].

12 McGinnis (2010), I, 55.

13 On the place of this premise in astronomical books, cfr. above note 11.

TABLE 2.1 Relations of mathematical sciences to physics

criterion → / science ↓	subject-matter	premises (I)	problems	premises (II)
arithmetic	+ (?)	-	-	-
geometry	+	-	-	-
moving sphere	+	+	-	-
music	+	+	-	-
optics	+	+	-	-
weights	+	+	-	-
astronomy	+	+	+	-

premises which express some essential properties of the Earth as a simple body, the astronomer bases his demonstration on propositions which are, in Avicenna's words, either "observational and optical" or geometrical.

The Arabic terms used by Avicenna to designate the non-geometrical kind of premises used in astronomy are *raṣadī* and *manāẓirī*. The first term designates not only the results of astronomical observations (*raṣad*), but also any kind of observational proposition. Almost the same is true of the second term (*manāẓirī*): it designates not only the propositions pertaining to the science of optics, but also any proposition that has something to do with the relative positions of the observer and the thing observed. So it is fair enough to translate both terms by "observational".

Avicenna's argument is summarised in table 2.1. As we can see, he uses the term "premise" in two different places and two different senses. What he means by this term in the beginning part of his discussion (column 3)—where he divides sciences into those who have some premises in common with physics and those who have not—is the "premises of a science" (*muqaddama ʿilm*), i.e., the most general principles on which a science is based. But in the last part of his exposition (column 5), where he talks about the propositions used by the astronomers to address a problem also addressed by the physicists, he uses the same term in a more restricted sense: what he means here by the term "premise" is the "of a syllogism" (*muqaddama qīyās*) i.e. each of the two propositions on which a syllogism is based.[14]

---

14 For these two meanings of the term "*muqaddama*", cfr. ʿAlī ibn Muḥammad al-Sharīf al-Jurjānī, *Kitāb al-Taʿrīfāt* (1985), 242. For *muqaddama* in the first sense, that is, the most

To put it briefly, if, in Avicenna's view, astronomy is distinct from any other mathematical science, it is because it employs in its demonstrations some propositions which are obtained through observation. These observational propositions intervene not only in the premises on which the whole science of astronomy is based,[15] but also, and above all, in many individual astronomical arguments.

To sum up, the text of Avicenna is an attempt at the demarcation between physics and the mathematical sciences as he understood them. He tries to locate the mathematical sciences in the general scheme of science, to define their place and to indicate their limitations. What emerges from this attempt is a hierarchical order, in which mathematical sciences are classified according to the degree of their closeness to, or remoteness from, physics. At the lowest rank we find arithmetic and geometry, while the highest rank is occupied by astronomy. Nevertheless, even astronomy, which, according to Avicenna, is the most physical among the mathematical sciences, cannot be considered as a full-fledged physical science. The reason is that it uses in its demonstrations some premises which are drawn from observation. This observational aspect, which is the strongest point of this science for a modern mind, is considered by Avicenna to be a disadvantage. The reason is that, because of this observational aspect, astronomical demonstrations cannot be causal or, as Avicenna puts it, in addressing the same problem, mathematical sciences give only the fact, whereas physics gives *both the fact and the why*. In other words, mathematical sciences proceed from effects to causes and not the other way round, so they lack one of the requirements, and in fact the most crucial requirement, of a real scientific explanation.

Avicenna's classification of mathematical sciences is in part descriptive, in the sense that it reflects the situation of these sciences in his time. It is in part prescriptive, because it determines the kind of propositions one can use in physics and in mathematical sciences. Avicenna warns the practitioners of both physics and astronomy against the mixing up of the two methods, i.e. he warns the physicist against basing his arguments on observational premises,

---

general principles on which a science is based, Avicenna usually uses the term "*mabādiʾ*" (singular: "*mabdaʾ*") which has a wider significance, including also the definitions of the basic terms used in a science.

15 Aristotelian scholars differ on the manner in which the premises on which a science should be based are acquired. At least some believe that they are acquired by induction, that is, by the generalisation of observational propositions, so an element of observation can intervene in every science. But as far as physics is concerned, this can happen only in the stage of the acquisition of principles; from then on, physics, in the Aristotelian sense, proceeds only through logical deduction from these principles.

because the use of this type of premises belongs to the domain of mathematical science of astronomy; he also warns the mathematician against using physical premises in his mathematical arguments.[16]

So, what emerges of this classification is two distinct groups of sciences. In one group we find all the mathematical sciences, whereas the other group has only one member: the astronomy. As we said earlier, the specificity of astronomy is that it treats of some problems also addressed by physics, but this is done by evoking a different type of propositions.

I think that this text of Avicenna is one of the rare occasions in Islamic philosophy where we witness a differentiation within the monolithic concept of science as a deductive structure, a differentiation based on the difference in kind between the propositions used, on the one hand, in astronomy and, on the other hand, in other mathematical sciences. Avicenna's advice to both physicists and the mathematicians is to not transgress their proper domains. Even in the problems which are common to astronomy and physics, the practitioners of each science have to proceed by their own methods. The observational method is proper to astronomy while the physicist is obliged to use only premises which concern the nature of the thing in question. With this division of labor, physics is denied any right to use premises which are obtained by observation and astronomy is denied any right to claim a truly scientific explanation.

The end result of Avicenna's text only accentuates a difficulty which was present from the very start in the Aristotelian conception of science. Many historians of science and mathematics have argued that the Aristotelian model for a demonstrative science, as a deductive structure based on a handful of premises, had its origin in the deductive structure of mathematical disciplines, later examples of which are found in Euclid (fl. 300 BCE)'s *Elements*, Euclid's *Optics*, and some Archimedean treatises. Moreover, Aristotle and his followers often cite mathematical arguments as examples of demonstrative argument. All this being said, the narrow Aristotelian conception of science, with its emphasis on the causal character of any demonstration, relegated the mathematical sciences to a lower rank, and so it created a clash between physics and the mathematical sciences which lasted from the Antiquity until the beginning of modern science. This conflict was overcome only by the triumph of modern

16 An example of this "mixing up" given by Avicenna is the following. If, to demonstrate that "the Earth is spherical" the astronomer argues that "the Earth is a simple body and every simple body should be spherical", he has presented a physical demonstration, and by consequence he has committed a methodological error. The same is true for the physicist who in his demonstration of the sphericity of the Earth, evokes the shape or the shadow by the Earth on the surface of the Moon [McGinnis (2010), I, 56].

science and the establishment of mathematical explanation as the sole type of legitimate scientific explanation in physics.

3 Physics and the Mathematical Sciences: A Historical Survey

Now I am going to present some aspects of this conflict in the science of the Islamic period. I begin by the more down-to-Earth aspects. The first thing to observe is the existence of a loose kind of division of labor between physicist-philosophers and mathematicians. As we will see later, the term "*aṣḥāb al-taʿālīm*" (mathematicians), frequently used both by the mathematicians and the others, referred to a distinct social group which had some overlapping with another social group consisting of philosophers, without being coextensive with it. Despite the fact that mathematics was considered as one of the three main divisions of philosophy, not all mathematicians were philosophers. In reality, some great mathematicians of this period (for example Thābit ibn Qurra and Naṣīr al-Dīn al-Ṭūsī) were equally great philosophers whereas some others were overtly anti-philosophical. An example of this type of mathematician is Samawʾal ibn Yaḥyā al-Maghribī, the great algebraist of the twelfth century, who was critical of many philosophical theses. Still others, like Abū Sahl al-Qūhī (and we can add, Ghiyāth al-Dīn al-Kāshī, mathematician and astronomer of the fifteenth century), were "at best indifferent to all forms of speculative thought".[17] The same is true for philosophers. Among them, some were very attentive to mathematics, while others completely ignored it. To sum up, mathematicians and physicist-philosophers were two different social groups, although in some cases the same person belonged to the two at the same time. The situation was not much different from the modern scientific landscape, in which some scientists have conscious philosophical engagements, others at most adopt a positivistic outlook and the rest do not pay any attention to philosophical foundations or implications of their work.

Moreover, not all the practitioners of the mathematical sciences were satisfied with their inferior rank in the Aristotelian scheme of science and their reaction took different forms. Some tried to show, by rhetorical devices, that mathematics is in fact superior to other branches of philosophy and some others tried to demonstrate the philosophical import of mathematical methods. Among this second group, some used mathematics to present a more precise formulation for a philosophical problem; others used mathematical methods

17 Rashed (2001), 159.

to put into question certain well-established physical theses; still others tried to show that mathematics can solve problems for which physicist cannot present any solution. In so doing, they were obliged sometimes to give a new interpretation for a philosophical concept and to appropriate some philosophical concepts to use them in their own way. In the remaining part of this article I try to elucidate these positions by presenting examples of each of these reactions.

3.1 *Rhetorical Arguments for the Superiority of the Mathematical Sciences*

This kind of argumentation goes back at least to Ptolemy (fl. second century A.D.), who, in the first pages of his *Almagest*, accepts the tripartite division of sciences, but argues that mathematical sciences are superior to the other two branches. The reason is that the subject-matter of theology is out of the reach of mortals, whereas the subject-matter of physics undergoes constant change and so cannot be the subject of true knowledge.

> For Aristotle divides theoretical philosophy too, very fittingly, into three primary categories, physics, mathematics and theology ... From all this we concluded: that the first two divisions of theoretical philosophy should rather be called guesswork than knowledge, theology because of its completely invisible and ungraspable nature, physics because of the unstable and unclear nature of matter; hence there is no hope that philosophers will ever agree about them; and that only mathematics can provide sure and unshakable knowledge to its devotees, provided one approaches it rigorously. For its kind of proof proceeds by indisputable methods, namely arithmetic and geometry.[18]

This kind of argument is found in the opening sentences of many mathematical works of the Islamic period. For example, in the beginning of his treatise on shooting stars, Abū Sahl Wayjan ibn Rustum al-Qūhī, mathematician of the tenth century, praises a group of scientists which "neither Galen nor anyone else could criticise—they could not criticise either them or their science, for they relied upon demonstrations in all their sciences and their books", and he adds that "they are the mathematicians".[19]

---

18 Ptolemy, *Almagest* (1998), 35.

19 Rashed (2001), 176.

In this kind of argument, the problem of causal explanation is generally avoided, and the emphasis is put on the certainty of the results obtained by mathematical reasoning. To use the words of Roshdi Rashed, "This praise of geometry and mathematics, reiterated throughout his [i.e. al-Qūhī's] writings, becomes the echo of an epistemic position that has an affinity with certain modern currents of thought".[20]

3.2 *Using Mathematics to Give a More Precise Answer to a Philosophical Problem*

An example of this kind of the use mathematics is Naṣīr al-Dīn al-Ṭūsī (1201–1274)'s treatise on the emanation of a plurality of beings from the One.[21] Even if the "mathematical" element of observation is absent from al-Ṭūsī's arguments, as Rashed has pointed out, al-Ṭūsī's mathematical treatment of this highly important philosophical problem has a double effect: it gives a quantitative answer to a problem hitherto treated only by speculative methods and, at the same time, it reduces the ontological content of the argument. Moreover, al-Ṭūsī's treatise is a big step in the foundation of the mathematical discipline of combinatorics.

3.3 *Using Mathematical Methods to Put into Question Some Well-Established Physical Theses*

A good example of this kind of argument is found in a treatise of Abū Sahl al-Qūhī, who tried to show, by using an observation, the falsity of the Peripatetic thesis according to which no infinite distance can be traversed in a finite time.[22] Al-Qūhī's argument is not without difficulty, because it implies a kind of begging the question.[23]

---

20 Rashed (2001), 160.

21 Rashed, "Combinatoire et métaphysique: Ibn Sīnā, al-Ṭūsī et al-Ḥalabī", in Rashed et Biard (eds.), (1999), 61–86.

22 Rashed, (1999), 7–24.

23 To refute the Aristotelian thesis, al-Qūhī considers a demi-circle opposite a source of light. When the demi-circle is traversed by the light ray emitted by the source, its image on the Earth is also traversed by the light ray. As in certain cases this image can be a branch of a hyperbola, and as this branch of hyperbola, which is infinite, is traversed by the light ray at the same time as the demi-circle (which is traversed in a finite time), so the Aristotelian thesis is refuted. The problem with this argument is that neither the movement of the light spot on the demi-circle nor the corresponding movement on the branch of hyperbola is the displacement of a body, while the Aristotelian thesis forbids only the infinite velocity of a *material body*.

Nevertheless, what is historically important is the intention of the author. As Rashed puts it: "this critic does not follow the method of the philosophers, with doctrinal emendations, but with a mathematical and experimental style".[24] The very idea that one can decide a physical question by using mathematical methods, that is, by evoking an observation, undermined a good part of the Peripatetic classification of sciences and the kind of demarcation we saw in Avicenna. In another work, which is known to us only through a treatise by the eleventh century Avicennian philosopher Abū al-ʿAbbās Lawkarī, the same al-Qūhī tried to adopt an Archimedean position and to refute an Aristotelian thesis pertaining to the concept of weight.[25]

3.4 *Mathematics Can Solve Problems for Which Philosophers Cannot Present Any Solution*

As we saw above, in Avicenna's classification of mathematical sciences astronomy was the only science which had problems in common with physics. In his scheme, the science of optics was less "physical" than astronomy and it occupied the same rank as music and the science of weights. This conception was partly due to the author's limited knowledge of this branch of mathematics, a knowledge which hardly went beyond the *Optics* of Euclid. Nevertheless, in the time Avicenna wrote his *Shifāʾ*, the science of optics was undergoing a profound transformation. The grounds for this change had already been prepared in the nineth and tenth centuries in the works of opticians such as al-Kindī (800–870), Ibn Sahl (fl. tenth century), Qustā ibn Lūqā (820/835–912), and others, but the full transformation took place with the emergence of the new optics of Ibn al-Haytham, known in Latin as Alhazen (965–1040).

The limits of time and space do not permit us to enter into the details of this transformation, suffice it to say that in this new optics experiment (which was not limited to observation and ordinary experience but included also experiments made with special devices designed by the author) occupied a central place. For the first time, "experimental proof" was a criterion which could decide any question pertaining to light and vision. The new optics of Ibn al-Haytham is expounded not only in his great work *The Book of Optics*, but also in several treatises in which he treats of specific optical problems. Among these problems, some, such as the halo and the rainbow, already belonged to the science of meteorology,[26] but others were totally new. In fact, the new Alhazenian

24 Rashed (1999), 7. For concrete examples of the method of the philosophers in treating certain mathematical problems, cfr. Rashed, (2016), 283–306.

25 Cfr. Masoumi Hamedani, "Al-Qūhī on the Concept of Weight", forthcoming.

26 A characteristic trait of this science was the relative instability of its subject matter, in the

conception of light was a tool by which problems that had never been subject to a scientific investigation could be reformulated in precise terms and put into a scientific analysis.

The best example for this kind of works is a group of treatises of Ibn al-Haytham, in each of which he addresses a physical problem never or seldom treated before as a problem which is in need of a scientific solution. This group includes: *On the Light of the Moon*, *On the Trace We See on the face of the Moon*, *On the Light of the stars*, and *On the Milky Way*.[27] Without entering into the details of these works, I only cite the introductory part of *On the Light of the Moon*, in which the author puts forward his program. In this treatise, Ibn al-Haytham follows a double aim: first, to demonstrate that the light of the Moon is borrowed from the Sun and, second, to show that the light of the Sun is *not* reflected upon the surface of the Moon. In other words, the Moon does not act like a mirror; it does not reflect the light according the laws of reflection, i.e. in certain directions and not in others. Rather, after receiving the light from the Sun, the Moon becomes a secondary source of light, and light is emitted from all the points of its surface and in all directions. He begins by saying that the idea that the Moon has no light of itself and that it receives its light from the Sun is accepted by all the researchers, then he goes on:

> All accomplished researchers agree on this point, despite the difference of their schools and the divergent opinions they hold on other scientific questions. Nevertheless, no one among them has presented a demonstrative discourse showing that this was necessarily the case. So, either no one among them has attempted to produce a demonstration—for its [presumed] evidence or for any other reason—or they had a demonstration for it, but their demonstration has not reached us. But, as far as a demon-

---

sense already expressed by Rashed: "that once we substitute rigorous methods in the place of uncontrolled observation and naive geometrical representation of the phenomenon (as with the research of Ibn al-Haytham on burning spheres, and that of Kamāl al-Dīn al-Fārisī on the model of the water-filled sphere), research on the rainbow became detached from the *Meteorologica* in order to become part of another domain: that of reformed optics", [Rashed (2001) 158. This point is also true of other phenomena treated by meteorology, including, as we will see, the Milky Way.

27 The Milky Way was a phenomenon on which mathematicians and physicist held different theories. The physicists, following Aristotle, believed that it is a meteorological phenomenon produced in the air, whereas the mathematicians held that it is part of the celestial world. To support their theory, the latter presented only an observational evidence: the absence of parallax. For a detailed discussion of this problem and the different positions of the two groups, cfr. Masoumi Hamedani, "La voie lactée: Alhazen et Averroès", in Ahmad Hasnawi (ed.) (2011), 39–62.

stration has not been presented showing that this is necessarily the case, a single thing is possible only: this is something contingent and not necessary, and it is conjectural and not certain. Among those who lived before us, no one has produced a satisfactory opinion on the quality of the light of this body [i.e. the Moon] after having received it from the Sun. As for the mathematicians, it seems that they believe that the body of the Moon has no light, that the light it sheds on the Earth is the solar ray which is reflected from its surface when it is illuminated by the Sun toward the Earth, that the body of the Moon is spherical, opaque, and polished, and when it is opposite the Sun and the solar ray ends up by arriving at its surface, the ray is reflected. Then [the reflected ray] continues [its trajectory] and whenever it encounters an opaque body, it lights it up, in the same way that the ray is reflected on mirrors and every polished body; that the luminous color the Moon presents when it is far from the Sun is the light of the Sun shed on it.

Nevertheless, on this subject no verified thesis has reached us from any of them, neither on how the Moon receives the light nor on how the light is reflected upon it.

As for those non-mathematicians who examine the nature of the heavenly bodies, they believe that the Moon receives its light from the Sun, and this is found among their sayings, but in a discursive and not a demonstrative way. As for the quality of the light it sheds on the Earth, they have not presented an opinion. All this is according to what we gathered from the sayings of the two groups which have reached us.

This being the case, and as we didn't find any sufficient saying investigating the truth about the quality of the light of this body, and as the humans search for the understanding the quality of the existing things and they don't come to rest except on arriving at a certainty which puts an end to all conjectures, this state of affairs incited us to investigate the quality of the light of this body, to exhaustively examine it and to unveil what was hidden about it.[28]

This text is highly revealing from different points of view. In the first place, the division of the scholars into mathematicians (*Aṣḥāb al-taʿālīm*) and the other researchers who are interested in the nature of the heavenly bodies, i.e. the physicist-philosophers, underlines not only the opposition of these two groups

28 For a critical edition of this text and its analysis, cfr. Masoumi Hamedani, *L'optique et la physique celeste: L'œuvre optico-cosmologique d'Ibn al-Haytham*, forthcoming; Masoumi Hamedani, "Ibn al-Hayṯam e la nuova fisica", in Rashed (ed.) (2002), III, 579–610.

but also the fact that the problem of the light of the Moon is among those investigated by these two, that is among the problems which are common to the mathematical sciences and physics. Seen in this light, Ibn al-Haytham's remark joins Avicenna's thesis discussed above.

In the second place, throughout this text the emphasis is put on the idea of demonstration and the certainty it brings about without mentioning the concept of cause. In the body of the treatise, Ibn al-Haytham presents only experimental and observational proofs for the theses he puts forward, and he never tries to complete his argument by bringing in "causal" arguments. It seems that in his view of the things, arguing from the effect, the observable phenomenon, to the cause is as decisive as arguing from the cause to the effect. All this is characteristic of the mathematicians' strategy, which replaces the "why" arguments with "how" arguments, without entering into a debate about their relative values and their respective places in the Aristotelian scheme of demonstration.

The examination of this group of treatises can shed some light on one of the most controversial theses put by him. At least on three occasions, Ibn al-Haytham talks of the necessity of combining the two methods of investigation, i.e. the method of mathematicians (*al-naẓar al-taᶜlīmī*) and the method of the physicists (*al-naẓar al-ṭabīᶜī*), in treating the problems pertaining to light and vision. This suggestion has given rise to a long debate among modern historians about the nature of this combination and the place of each of these components in the final result of investigation. The least we can say is that what Ibn al-Haytham meant by this "combination" cannot be reduced to an adding up of two doctrines, one mathematical and the other physical, about the nature of light, or to an investigation of which the matter is provided by physics (in the Aristotelian sense) while the form is mathematical. To understand the real significance of this kind of "combination", the best thing to do is to examine Ibn al-Haytham's own writings and the way he follows in his research. Here we witness that "physical" premises, those who talk about the nature of things and their essential properties, are seldom mentioned. Instead, the emphasis is put on "mathematical" method of investigation, a method that proceeds by observation and experimentation, while the results obtained by using this method are presented as being certain and necessary.

As noted above, for a peripatetic philosopher like Avicenna, who didn't know Ibn al-Haytham, combining the physical and mathematical methods was not legitimate and he warned against such an attempt. Averroes, another great Aristotelian philosopher and commentator, follows the same line where, in his Middle Commentary on Aristotle's *Meteorology*, he condemns it, saying that to mix up the two methods, as Ibn al-Haytham has done, is an error (*wa man jamaʿa al-naẓarayn, faqad akhṭaʾa kamā faʿala Ibn al-Haytham*), while in his

Short Commentary on the same book, written before the Middle Commentary, he had already tried to somehow justify this kind of "combination".

This change of position can be explained by the fact that when he wrote his Shorter Commentary, Averroes didn't know an example of what Ibn al-Haytham meant by the combination of the two methods, whereas, in his Middle Commentary, he shows a certain knowledge of the latter's work. He explicitly cites Ibn al-Haytham's *On the Light of the Moon*, he accepts the main result obtained in this treatise (that the light of the Sun is not reflected on the surface of the Moon), and curiously enough, he rejects the method proposed by Ibn al-Haytham. In fact, what urges Averroes to disagree with Ibn al-Haytham is not his ignorance of the writings of the latter, but his relative knowledge of the way this combination is done and the extent to which it would affect the peripatetic concept of science.

3.5 *Appropriation of a Philosophical Concept for Using or Abusing It*

In their treatment of the problems which were common to physics and a mathematical science, the mathematicians sometimes used not only experimental and mathematical tools but also concepts already elaborated by the philosophers. Nevertheless, this necessitated certain changes which the mathematician deemed necessary for the object he followed.

The best example we know of this kind of mathematicians' strategy is the use Ibn al-Haytham makes, throughout his work and especially in his *Optics*, of the concept of form. One peculiarity of this concept, as it is used by Ibn al-Haytham, is that the form of an object, in the Alhazenian sense, is no longer an indivisible property belonging to the object as a whole, but the sum total of the forms of its constituent parts or, more precisely, of the smallest parts of that object that can have a form. This semantic change was not arbitrary, but a development necessitated by the history of optics at least from al-Kindī (800–870) onwards.

4 Conclusion

The Aristotelian tripartite division of theoretical philosophy into physics, mathematics and metaphysics has been very influential throughout the Antiquity and the Middle Ages. This division is often thought to have two main consequences: on the one hand, a subordination of mathematical and physical sciences to metaphysics and, on the other hand, the existence of a unique method used in all the three groups of sciences.

Nevertheless, the existence of a class of problems dealt with by both physics and mathematical sciences is attested from the time of Aristotle himself. In Aristotle, this situation gives rise to a tacit classification of mathematical sciences according to their closeness to or remoteness from physics. This line of thought is followed by Avicenna in his *Shifāʾ*, where he tries to give a more detailed account of the relations between physical and mathematical sciences. Although the Avicennian classification implies a justification of the observational methods of the mathematicians, it denied them any right to produce a true, i.e. causal, explanation. For the problems which were examined by both mathematicians and the physicists, the true explanation remained always the one presented by the physicist and, to use Avicenna's words, was based on "what was necessitated by the natural body as natural body".

I hope to have shown that not everybody was pleased by this division of labor, and that the history of the relations between physics and the mathematical sciences in Islam is rather a history of explicit or implicit conflict. The study of the forms this conflict took can shed some light not only on the history of science in Islam but on the history of medieval and early modern science as well.

Bibliography

Aḥmad ibn Kathīr al-Farghānī (ninth century), *Jawāmiʿ ʿilm al-nujūm* (MS 967H, Princeton: Garrett, fol. 10a–16a).

Alī ibn Muḥammad al-Sharīf al-Jurjānī (1340–1413), *Kitāb al-Taʿrīfāt* (Beirut: Maktabat Lubnān, 1985).

Aristotle (384–322 BCE), *Physics*, Books I–II, transl., Introduction and Notes W. Charlton (Oxford: Oxford University Press, 1970).

Aristotle, *Meteorology*, in *The Complete Works of Aristotle*, ed. Jonathan Barnes, the revised Oxford Translation, Vols. I–II (Princeton New Jersey: Princeton University Press, 1984), 554–625.

Duhem, Pierre, *Le système du monde. Histoire des doctrines cosmologiques de Platon à Copernic*, Vols. I–X (Paris: Librairie scientifique A. Hermann et fils, 1913–1959).

Granger, Gilles Gaston, *La théorie aristotélicienne de la science* (Paris: Aubier, 1976).

Ibn Sinā (980–1037), *Al-samāʿ al-ṭabīʿī*, transl. Jon McGinnis, *The Physics of The Healing*, Vols. I–II, parallel English-Arabic text (Provo, Utah: Brigham Young University Press, 2010).

Ibn Sinā, *Aqsām ʿulūm al-awāʾil* (MS 712, Tehran: Majlis, fol. 278).

Kadivar, Mohsin, "Ibn Sīnā wa ṭabaqi-bandī-yi ḥikmat, taḥlīl, taḥqīq wa taṣḥiḥ Risāla *Aqsām al-Ḥikma*", *Jāwīdān khirad* [*Sophia Perennis*], 5: 1 (winter 2009), 111–137 [in Persian].

Kūshyār ibn Labbān al-Jīlī (971–1029), *Al-Zīj al-jāmiʿ* (MS Yeni Cami 784, Istanbul: Süleymaniye Kütüphanesi, fol. 319b–320b).

Masoumi Hamedani, Hossein, "Ibn al-Hayṯam e la nuova fisica", in Roshdi Rashed (ed.), *Storia della Scienza: vol. III, La civilta islamica* (Roma: Instituto della Enciclopedia Italiana, 2002), 579–610.

Naṣīr al-Dīn al-Ṭūsī (1201–1274), *Al-Risāla al-Muʿīniyya*, ed. Hasan Amini, MA thesis, University of Tehran, 1387/2008.

Naṣīr al-Dīn al-Ṭūsī, *Al-Tadhkira fī ʿilm al-hayʾa*, ed. Jamil Ragep, *Naṣīr al-Dīn al-Ṭūsī's Memoir On Astronomy* Vols. 1-II (New York: Springer, 1993), I, 90–107.

Ptolemy (100–168 A.D.), *Optics*, transl. A. Mark Smith, *Ptolemy's Theory of Visual Perception. An English Translation of the Optics with Introduction and Commentary* (Philadelphia: American Philosophical Society, 1996).

Ptolemy, *Almagest*, transl. G.J. Toomer (Princeton New Jersey: Princeton University Press, 1998).

Rashed, Roshdi, "Al-Qūhī: From Meteorology to Astronomy", *Arabic Sciences and Philosophy*, 11: 2 (2001), 157–204.

Rashed, Roshdi, "Combinatoire et métaphysique: Ibn Sīnā, al-Ṭūsī et al-Ḥalabī", in Rashed, Roshdi & Biard, Joël (éds.), *Les Doctrines de la science de l'antiquité à l'âge classique*, Leuven, Peeters, 1999, pp. 61–86.

Rashed, Roshdi, "Combinatoire et métaphysique: Ibn Sīnā, al-Ṭūsī et al-Ḥalabī", in

Rashed, Roshdi, "Al-Qūhī Vs. Aristotle on Motion", *Arabic Sciences and Philosophy*, 9: 1 (March 1999), 7–24.

Rashed, Roshdi "Avicenne, 'philosophe analytique' des mathématiques", *Les études philosophiques*, 2 (Avril 2016), 283–306.

Sajjad Nikfahm-Khubravan, Osama Eshera, "The Five Arabic Revisions of Autolycus' *On the Moving Sphere* (Proposition VII)", *Tarikh-e Elm*, 16: 2 (2019), 7–71.

Simon, Gerard, *L'être, le regard et l'apparence dans l'optique de l'antiquité* (Paris: Seuil, 1989).

CHAPTER 3

Ibn al-Haytham: between Mathematics and Physics

Roshdi Rashed

For the vast majority of historians, and, more generally, of laymen, Ibn al-Haytham's (965–1040 ca.) major contribution concerns the vision in all its aspects (physical, physiological and psychological) and, namely, the causes of perceptual and cognitive effects. The objective of Ibn al-Haytham, according to them, was mainly to abandon the traditional theory of vision, and to replace it by a new one. In spite of this reform he would belong to both ancient and medieval traditions in so far as he was concerned with vision and sight.

I will argue here that this reform was a minor consequence of a more general and more fundamental research programme, and even his conception of the science of optics is quite different as so far that his main task was about light, its fundamental properties and how they determine its physical behaviour, as reflection, refraction, focalisation, etc.

Some historians of optics consider that, up to the seventeenth century in Europe, the science of optics before Kepler (1571–1630) was aimed primarily at explaining vision. The merest glance at the optical works of Ibn al-Haytham leaves no doubt that this global judgment is far from being correct. Indeed, this statement is correct as far as it concerns the history of optics before the shift done by Ibn al-Haytham and the reform he accomplished. Successor of Ptolemy (fl. second century A.D.), al-Kindī (800–870) and Ibn Sahl (fl. tenth century), to mention only a few, he unified the different branches of optics: optics, dioptrics, anaclastics, meteorological optics, etc. This unification was possible only for a mathematician who focused on light, and not on vision. Nobody, as far as I know, before Ibn al-Haytham, wrote such books titled: *On light*; *On the light of the moon*; *On the light of the stars*; *On the shadows*, among others, in which nothing concerns sight. At the same time, three books from his famous *Book of Optics* are devoted strictly to the theory of light. None of the authors before him, who were mainly interested in vision, wrote a very important contribution on physical optics such as the one on *The Burning Sphere*.

I begin by quoting the expression which Ibn al-Haytham repeated more than once in his different writings on optics. At the beginning of this famous *Book of Optics*, he writes:

 | DOI:10.1163/9789004513402_005

> Our subject is obscure and the way leading to knowledge of its nature difficult, moreover our inquiry requires a combination of the natural and mathematical sciences.[1]

But such a combination in optics, for instance, requires one to examine the entire foundations and to invent the means and the procedures to apply mathematics on the ideas of natural phenomena. For Ibn al-Haytham, it was the only way to obtain a rigorous body of knowledge.

Why this particular turn, at that time? Let me remind that Ibn al-Haytham lived at the turn of the first millennium. He was the heir of two centuries of scientific research and scientific translations, in mathematics, in astronomy, in statics, in optics, etc. His time saw intense research in all these fields. He himself wrote more in the fields of mathematics and astronomy than in the field of optics *per se*. According to early biobibliographers, Ibn al-Haytham wrote twenty-five astronomical works: twice as many works on the subject as he did in optics. The number of his writings alone indicates the huge size of the task accomplished by him and the importance of astronomy in his life's work. In all branches of mathematics, he wrote more than all his writings in astronomy and in optics together. He wrote in optics the famous huge book, *Kitāb al-Manāẓir—The book of Optics—*, in astronomy likewise he wrote a huge book entitled *The configuration of the motions of each of the seven wandering stars*.

Before coming back in some details to these contributions, let me characterise Ibn al-Haytham's research programme.

1. It is a new one, concerning the relationships between mathematics and natural phenomena, never conceived before. His aim is to mathematise every empirical science. This application of mathematics can take different forms, not only given to the different disciplines, but also in one and the same discipline.
2. It does not concern only optics, but every natural science, i.e., for the epoch, astronomy and statics.
3. Its success depends on the means—mathematical, linguistic and technical—by which mathematics control the semantic and syntactical structures of natural phenomena.

I. To put the facts right, I will turn at first, quite briefly, to Ibn al-Haytham's astronomy. He wrote at least three books criticizing the astronomical theory of Ptolemy:

1 Ibn al-Haytham, *Kitāb al-manāẓir* (1989), I, 4.

1. *The Doubts concerning Ptolemy*
2. *Corrections to the* Almagest
3. *The Resolution of Doubts concerning the* Almagest

In the *Doubts*, Ibn al-Haytham comes to the conclusion that "the configuration Ptolemy assumes for the motions of the five planets is a false one".[2] A few lines further on, he continues: "The order in which Ptolemy had placed the motions of the five planets conflicts with the theory ⟨that he had proposed⟩".[3] A little later, he states: "The configurations that Ptolemy assumed for the ⟨motions of⟩ the five planets are false ones. He decided on them knowing they were false, because he was unable ⟨to propose⟩ other ones."[4] After such comments, and many others like them in several places of his writings, Ibn al-Haytham had no option but to construct a planetary theory of his own, on a solid mathematical basis, and free from the internal contradictions found in Ptolemy's *Almagest*. For this purpose, he conceived the idea of writing his monumental and fundamental book *The Configuration of the Motions of the Seven Wandering Stars*. If we wish to characterise the irreducible inconsistencies that, according to Ibn al-Haytham, vitiate Ptolemy's astronomy, we may say that they arise from the poor fit between a mathematical theory of the planets and a cosmology; that is, the combination between mathematics and physics. Ibn al-Haytham was familiar with similar, though of course not identical, situations when, in optics, as we shall see, he encountered the inconsistency between geometrical optics and physical optics as understood not only by Euclid and Ptolemy, but also by Aristotle and the philosophers.

In *The Configuration of the Motions* he deals with the apparent motions of the planets, without ever raising the question of the physical explanation of these motions in terms of dynamics. It is not the causes of celestial motions that interest Ibn al-Haytham, but only the motions themselves observed in space and time. Thus, to proceed with the systematic mathematical treatment, and to avoid the obstacles that Ptolemy had encountered, he first needed to break away from any kind of cosmology. Thus the purpose of Ibn al-Haytham's *Configuration of the Motions* is clear: instead of constructing, as his predecessors, a cosmology, or a kind of dynamics, he constructs the first geometrical kinematics.

A close examination of the way he organises his exposition of planetary theory shows that Ibn al-Haytham begins by omitting physical spheres and by proposing simple—in effect, descriptive—models of the motions of each of the

2 Rashed (2014), 13.
3 Ibid.
4 Ibid.

seven planets. As the exposition progresses, he makes the models more complicated and increasingly subordinates them to the discipline of mathematics. This growing mathematisation leads him to regroup the motions of several planets under a single model. This step obviously has the effect of privileging a property that is common to several motions. In this way Ibn al-Haytham opens up the way to achieving his principal objective: to establish a system of celestial kinematics. He does so without as yet formulating the concept of instantaneous speed, but by using the concept of mean speed, represented by a ratio of arcs.

In the course of his research, which I analysed elsewhere,[5] we encounter a concept of astronomy that is new in several respects. Ibn al-Haytham sets himself the task of describing the motions of the planets exactly in accordance with the paths they draw on the celestial sphere. He is neither trying 'to save the phenomena', like Ptolemy, that is, to explain the irregularities in the assumed motion by means of artifices such as the equant; nor trying to account for the observed motions by appealing to underlying mechanisms or hidden natures. He wants to give a rigorously exact description of the observed motions in terms of mathematics. Thus his theory for the motion of the planets calls upon no more than observation and conceptual constructs susceptible to explaining the data, such as the eccentric circle and in some cases the epicycle. However, this theory does not aim to describe anything beyond observation and these concepts, and in no way is it concerned to propose a causal explanation of the motions.

The new astronomy no longer aims at constructing a model of the universe, as in the *Almagest*, but only at describing the apparent motion of each planet, a motion composed of elementary motions, and, for the inferior planets, also of an epicycle. Ibn al-Haytham considers various properties of this apparent motion: localisation and the kinematic properties of the variations in speed.

In this new astronomy, as in the old one, every observed motion is circular and uniform, or composed of circular and uniform motions. To find these motions, Ibn al-Haytham uses various systems of spherical coordinates: equatorial coordinates (the required time and its proper inclination); horizon coordinates (altitude and azimuth); and ecliptic coordinates. The use of equatorial coordinates as a primary system of reference marks a break with Hellenistic astronomy. In the latter, the motion of the orbs was measured against the

5 Rashed (2014).

ecliptic, and all coordinates were ecliptic ones (latitude and longitude). Thus, basing the analysis of the planets' motion on their apparent motions drives a change in the reference system for the data; we are now dealing with right ascension and declination. Ibn al-Haytham's book thus transports us into a different system of analysis.

To sum up, in the *The Configuration of the Motions*, Ibn al-Haytham's purpose is purely kinematics; more precisely, he wanted to lay the foundations of a completely geometrical kinematics tradition. But carrying out such a project involves first of all developing some branches of geometry, as also of plane and spherical trigonometry. In both fields, Ibn al-Haytham obtained new and important results.

In astronomy, properly, there are two major processes that are jointly involved in carrying through this project: freeing celestial kinematics from cosmological connections, that is, from all considerations of dynamics, in the ancient sense of the term; and to reduce physical entities to geometrical ones. The centres of the motions are geometrical points without physical significance; the centres to which speeds are referred are also geometrical points without physical significance; even more radically, all that remains of physical time is the 'required time', that is, a geometrical magnitude. In short, in this new kinematics, we are concerned with nothing that identifies celestial bodies as physical bodies. All in all, though it is not yet that of Kepler, this new kinematics is no longer that of Ptolemy nor of any of Ibn al-Haytham's predecessors; it is *sui generis*, halfway between Ptolemy and Kepler. It shares two important ideas with ancient kinematics: every celestial motion is composed of elementary uniform circular motions, and the centre of observation is the same as the centres of the Universe. On the other hand, it has in common with modern kinematics the fact that the physical centres of motions and speeds are replaced by geometrical centres.

In fact, once Ibn al-Haytham had engaged upon mathematising astronomy and had noted not only the internal contradictions in Ptolemy, but doubtless also the difficulty of constructing a self-consistent mathematical theory of material spheres using an Aristotelian physics, he conceived the project of giving a completely geometrised kinematic account.

Ibn al-Haytham had the same experience in optics. In astronomy, kinematics and cosmology are entirely separated in order to accomplish a reform of the discipline. In the same way, in optics work on light and its propagation is entirely separated from work on vision in order to accomplish a reform of optics; in the one case as in the other, we shall see, Ibn al-Haytham arrived at a new idea of the science concerned.

II. Let us come now to Ibn al-Haytham's optics. As we have said above, Ibn al-Haytham was preceded by two centuries of translation into Arabic of the main Greek optical writings, as well of inventive research. Among his Arabic predecessors were al-Kindī, Qusṭā ibn Lūqā (between 820 and 835–912), Aḥmad ibn ʿĪsā ʿUṭārid, etc. During these two centuries, the interest shown in the study of burning mirrors is an essential part of the comprehension of the development of catoptrics, anaclastics and dioptrics, as the book written between 983–985 by the mathematician al-ʿAlāʾ ibn Sahl (940–1000) testifies. Before this contribution of Ibn Sahl, the catoptricians like Diocles (fl. end third–beginning fourth century BCE), Anthemius of Tralles (474–574 A.D.) al-Kindī etc.[6] asked themselves about geometrical properties of mirrors and about light they reflect at a given distance. Ibn Sahl modifies the question by considering not only mirrors but burning instruments, i.e. those which are susceptible to light not only by reflection, but also by refraction; and how in each case the focalisation of light is obtained. Ibn Sahl studies then, according to the distance of the source (finite or infinite) and the type of lighting (reflection or refraction) the parabolic mirror, the ellipsoidal mirror, the plano-convex lens and the biconvex lens. In each of these, he proceeds to a mathematical study of the curve, and, then, expounds a mechanical continuous drawing of it. For the plano-convex lens, for instance, he starts by studying the hyperbola as a conic section, in order then to take up again a study of the tangent plane to the surface engendered by the rotation of the arc of hyperbola around a fixed straight line, and, finally, the curve as an anaclastic curve, and the laws of refraction.

These studies focused on light and its physical behaviour were instrumental in the discovery by Ibn Sahl of the concept of a constant ratio, characteristic of the medium, which is a masterpiece in his study of refraction in lenses, as well as his discovery of the so-called Snellius' law.

Thus, Ibn Sahl had conceived and put together an area of research into burning instruments and, also, anaclastics. But, obliged to think about conical figures other than the parabola and the ellipse—the hyperbola for example—as anaclastic curves, he was quite naturally led to the discovery of the law of Snellius.

Rich in technical material, this new discipline is in fact very poor on physical content: it is faint and reduces a few energy considerations. By way of example, at least in his writings that have reached us, Ibn Sahl never tried to explain why certain rays change direction and are focused when they change medium: it is enough for him to know that a beam of rays parallel to the axis

6 Rashed (2000); Rashed (1997).

of a plano-convex hyperbolic lens gives by refraction a converging beam. As for the question why the focusing produces a blaze, Ibn Sahl is satisfied with a definition of the luminous ray by its action of setting ablaze by postulating, as did his successors elsewhere for much longer, that the heating is proportional to the number of rays.

Whilst Ibn Sahl was finishing his treatise on *Burning Instruments* very probably in Baghdad, Ibn al-Haytham was probably beginning his scientific career.

Compared with the writings of the Greek and Arab mathematicians who preceded him, the optical work by Ibn al-Haytham presents at first glance two striking features: extension and reform. It will be concluded on a more careful examination that the first trait is the material trace of the second. In fact, no one before Ibn al-Haytham had embraced so many domains in his research, collecting fairly independent traditions: mathematical, philosophical, medical. The titles of his books serve moreover to illustrate this large spectrum: *The Light of the Moon, The Light of the Stars, The Rainbow and the Halo, Spherical Burning Mirrors, Parabolical Burning Mirrors, The Burning Sphere, The Shape of the Eclipse, The Formation of Shadows, On Light*, as well as his *Book of Optics* translated into Latin in the twelfth century and studied and commented on in Arabic and Latin until the seventeenth century. Ibn al-Haytham therefore started not only the traditional themes of optical research but also others, new ones, to cover finally the following areas: optics, catoptrics, dioptrics, physical optics, meteorological optics, burning mirrors, the burning sphere.

A more careful look reveals that, in the majority of these writings, Ibn al-Haytham pursued the realisation of his programme to reform the discipline by taking up every different problem in turn. The founding action of this reform consisted in making clear the distinction, for the first time in the history of optics, between the conditions of propagation of light and the conditions of vision of objects. It led on one hand to providing physical support for the rules of propagation—it concerns a mathematically guaranteed analogy between a mechanical model of the movement of a solid ball thrown against an obstacle, and that of the light—and, on the other hand, to proceeding everywhere geometrically and by observation and experimentation. It led also to the definition of the concept of light ray and light bundle as a set of straight lines on which light propagates, rays independent from each other which propagate in an homogeneous region of space. These rays are not modified by other rays which propagate in the same region.

Thanks to the concept of light bundle, Ibn al-Haytham was able to study the propagation and diffusion of light mathematically and experimentally. Optics no longer has the meaning that is assumed formerly: a geometry of perception. It includes henceforth two parts: a theory of vision, with which are also associated a physiology of the eye and a psychology of perception, and a theory of light, to which are linked geometrical optics and physical optics. The organisation of his *Book of Optics* reflects already the new situation. In it are books devoted in full to propagation—the third chapter of the first book and Books IV to VII; others deal with vision and related problems. This reform led to, amongst other things, the emergence of new problems, never previously posed, such as the famous "problem of Alhazen" on catoptrics, the examination of the spherical lens and the spherical dioptre, not only as burning instruments but as optical instruments, in dioptrics; and to experimental control as a practice of investigation as well as the norm for proofs in optics and more generally in physics.

Let us follow now the realisation of his reform in the *Book of Optics* and in other treatises. This book opens with a rejection and a reformulation. Ibn al-Haytham rejects straightaway all the variants of the doctrine on the visual ray, to ally himself with philosophers who defended a doctrine of intromission on the form of visible objects. A fundamental difference remains nevertheless between him and the philosophers, such as his contemporary Avicenna: Ibn al-Haytham did not consider the forms perceived by the eyes as "totalities" which radiate from the visible object under the effect of light, but as reducible to their elements: from every point of the visible object radiate a ray towards the eye. The latter has become without soul, without πνεῦμα ὀπτικόν, a simple optical instrument. The whole problem was then to explain how the eye perceives the visible object with the aid of these rays emitted from every visible point.

After a short introductory chapter, Ibn al-Haytham devotes two successive chapters—the second and the third books of his *Book of Optics*—to the foundations of the new structure. In one, he defines the conditions for the possibility of vision, while the other is about the conditions for the possibility of light and its propagation. These conditions, which Ibn al-Haytham presents in the two cases of empirical notions, i.e. as resulting from an ordered observation or a controlled experiment, are effectively constraints on the elaboration of the theory of vision, and in this way on the new style of optics.

The conditions for vision detailed by Ibn al-Haytham are six: the visible object must be luminous by itself or illuminated by another; it must be opposite the eye, i.e. one can draw a straight line to the eye from each of its points; the medium that separates it from the eye must be transparent, without being cut into by any opaque obstacle; the visible object must be more opaque than this medium; it must be of a certain volume, in relation to the visual sharpness. These are the notions, writes Ibn al-Haytham, "without which vision cannot take place". These conditions, one cannot fail to notice, do not refer, as in the ancient optics, to those of light or its propagation. Of these, the most important, established by Ibn al-Haytham, are the following: light exists independently of vision and exterior to it; it moves with great speed and not instantaneously; it loses intensity as it moves away from the source; the light from a luminous source—substantial—and that from an illuminated object—second or accidental—propagate onto bodies which surround them, penetrate transparent media, and light up opaque bodies which in turn emit light; the light propagates from every point of the luminous or illuminated object in straight lines in transparent media and in all directions; these virtual straight lines along which light propagates form with it "the rays"; these lines can be parallel or cross one another, but the light does not mix in either case; the reflected or refracted light propagates along straight lines in particular directions. As can be noted, none of these notions relate to vision.

A theory of vision must henceforth answer not only the six conditions of vision, but also the conditions of light and its propagation. Ibn al-Haytham devotes the rest of the first book of his *Book of Optics* and the two following books to the elaboration of this theory, where he takes up again the physiology of the eye and a psychology of perception as an integral part of this new theory of intromission. Three books of the *Book of Optics*—the fourth to the sixth—deal with catoptrics.

Let us consider some aspect of this research into catoptrics by Ibn al-Haytham. He restates the law of reflection, and explains it with the help of the mechanical model already mentioned. Then he studies this law for different mirrors: plane, spherical, cylindrical and conical. In each case, he applies himself above all to the determination of the tangent plane to the surface of the mirror at the point of incidence, in order to determine the plane perpendicular to this last plane, which includes the incident ray, the reflected ray and the normal to the point of incidence. Here as in his other studies, to prove these results experimentally, he conceives of and builds an apparatus inspired by the one that Ptolemy constructed to study reflection, but more complicated and adaptable to every case. Ibn al-Haytham also studies the image of an object and its position in the different mirrors. He applies himself to a whole class of problems: the determination of the incidence of a given reflection in the different mirrors and conversely. He also poses for the different mirrors the problem which his name is associated: given any two points in front of a mirror, how does one determine on the surface of the mirror a point such that the straight line which joins the point to one of the two given points is the incident ray, whilst the straight line that joins this point to the other given point is the reflected ray. This problem, which rapidly becomes more complicated, has been solved by Ibn al-Haytham.

Ibn al-Haytham pursues this catoptric research in other essays, some of which are later than the *Book of Optics*, such as *Spherical Burning Mirrors*.[7] It is in this essay of a particular interest that Ibn al-Haytham discovers the longitudinal spherical aberration.

Ibn al-Haytham devotes a substantial part of the seventh book to the study of the image of an object by refraction, notably if the surface of separation of the two media is either plane or spherical. It is in the course of this study that he settles on the spherical dioptre and the spherical lens, following thus in some way the research by Ibn Sahl, but modifying it considerably; this study of the dioptre and the lens appears in effect in the chapter devoted to the problem of the image, and is not separated from the problem of vision. For the dioptre, Ibn al-Haytham considers two cases, depending on whether the source—punctual and at a finite distance—is found on the concave or convex side of the spherical surface of the dioptre.

Ibn al-Haytham studies the spherical lens, giving particular attention to the image that it gives of an object. He restricts himself nevertheless to the examination of only one case, when the eye and the object are on the same diameter. Put another way, he studies the image through a spherical lens of an object placed in a particular position on the diameter passing through the eye. His procedure is not without similarities to that of Ibn Sahl when he studied the biconvex hyperbolic lens. Ibn al-Haytham considers two dioptres separately, and applies the results obtained previously. It is in the course of his study of the spherical lens that Ibn al-Haytham returns to the spherical aberration of a point at a finite distance in the case of the dioptre, in order to study the image of a segment which is a portion of the segment defined by the spherical aberration.

Let us stop at this point on spherical aberration, to conclude.

With Ibn al-Haytham, one result has been definitively obtained: the half century which separates him from Ibn Sahl should be counted among the distinctive moments in the history of optics: dioptrics appears to have extended its domain of validity and, by its very progress, to have changed its orientation. With Ibn al-Haytham, the conception of dioptrics as a geometry of lenses has become outdated. Here again, in his own words, we must combine mathematics and physics in order to study dioptres and lenses, whether burning or not. The mathematisation could only be achieved with Ibn al-Haytham because he

7 Ibn al-Haytham (Berlin, Oct. 2970/7).

separated the study of the natural phenomenon of light from this of vision and sight. In optics as in astronomy the research programme of Ibn al-Haytham is the same: mathematise the discipline and combine this mathematisation with the ideas of the natural phenomena.

Bibliography

Ibn al-Haytham (965–1040), *Fī al-marāyā al-muḥriqa bi-al-dawā'ir* (MS Oct. 2970/7, Berlin: Staatsbibliotheek), fols. 66r–73v.

Ibn al-Haytham, *Kitāb al-manāẓir*, transl. A.I. Sabra, *The Optics of Ibn al-Haytham*, Books 1–3 (London: the Warburg Institute, 1989).

Rashed, Roshdi, *Ibn al-Haytham. New Astronomy and Spherical Geometry*, A History of Arabic Sciences and Mathematics, vol. 4 (London: Routledge, 2014).

Rashed, Roshdi, *Les Catoptriciens grecs*, vol. 1: *Les miroirs ardents*, édition, traduction et commentaire par Roshdi Rashed (Paris: Les Belles Lettres, 2000).

Rashed, Roshdi, *Œuvres philosophiques et scientifiques d'al-Kindī*, vol. 1: *L'Optique et la Catoptrique d'al-Kindī* (Leiden: E.J. Brill, 1997) [Arabic translation: *'Ilm al-manāẓir wa-'ilm in'ikās al-ḍaw'*, Silsilat Tārīkh al-'ulūm 'inda al-'Arab 6 (Beyrouth: Markaz Dirāsat al-Waḥda al-'Arabiyya, 2003)].

CHAPTER 4

La Musique parmi les sciences dans les textes arabes médiévaux

Anas Ghrab

L'univers des sons, et en particulier celui de la musique, est l'une des manifestations anthropologiques les plus importantes qui ont profondément affecté la pensée humaine. Cela peut s'expliquer par les multiples manifestations du champ sonore et par sa présence continue tout au long de la vie humaine quotidienne. En même temps, le phénomène purement acoustique et sonore, constitue un objet invisible et immatériel, tout en étant lié à des objets physiques et à de la matière. Ces objets constituent les instruments qui génèrent les sons, qu'ils soient naturels (sons de la nature, de la gorge humaine, etc.) ou artificiels (idiophones, aérophones, cordophones, etc.). De ce fait le son peut être entendu sans être vu, tout en ayant un impact sur l'auditeur. Ainsi, certainement dès les débuts de la civilisation humaine, le son en tant que tel a attiré l'attention des hommes de science et de la religion quelles que soient leur vision du monde.

C'est dans ce cadre que certains auteurs arabes, dans la continuité de traditions antérieurs, se sont intéressés au son, à la musique et aux concepts liés à ce domaine. Les premiers textes théoriques en arabe sur la musique remontent au IXe siècle, et progressivement, un concept de « Science de la Musique » (*ʿIlm al-Mūsīqā*) a émergé parallèlement à la vague de traduction des textes grecs.

Certains travaux ont tenté d'étudier cet aspect de la musique[1], mais malheureusement, leur approche présente principalement des aspects secondaires de cette thématique, liés au positionnement des ligatures sur le *ʿŪd*, sans aborder de ce qui justifie la scientificité recherchée dans ces textes. En complément, nous allons alors synthétiser dans ce qui suit l'évolution de ce concept depuis les premiers textes jusqu'au XVe siècle environ, moment où les textes cherchent de moins en moins à se référer à une attitude « scientifique ». Nous verrons tout particulièrement comment l'idée de Science de la Musique évolue d'une « science philosophique », l'Harmonique, vers une science indépendante, en

1 Nous pensons ici particulièrement à l'article de Chabrier (1997), 231-262.

 | DOI:10.1163/9789004513402_006

passant par le statut de science mathématique, et nous observerons tout particulièrement les réflexions qui ont été menées concernant son rapport à la physique (science de la Nature).

1 Science de l'Harmonique

La gnomologie greco-arabe des débuts du IXe siècle constitue le cadre privilégié où le terme « musique » (*musīqā*) a commencé à faire son introduction dans la langue arabe. Les propos des Philosophes (*Adāb* et *Nawādir al-falāsifa*) relayés par Būlos (n.d.)[2], ʿAlī Ibn ʿUbayda al-Rayḥānī (ob. 834) ainsi que Ḥunayn Ibn Isḥāq (808-873)[3], ont été repris par la suite par al-Kindī (800–870) et par les Ikhwān al-Ṣafāʾ (xe siècle). Cette collection de propos montre en particulier une conscience que les sons, y compris la musique, avaient un effet sur l'âme. Ce constat, simple en apparence, aura un impact majeur sur les théories qui en dérivent, qui sont elles-mêmes dans la tradition néo-platonicienne, comme nous allons le voir.

Ainsi, malgré l'émergence du terme « musique » dans la langue arabe et la rédaction de textes liés à la théorie musicale proprement-dite, tels que *Kitāb al-Nagham*[4] de Yaḥyā Ibn ʿAlī Ibn al-Munajjim d. 912, ou dans sa relation avec la pratique musicale telle que la *Question de la musique*[5] de Thābit Ibn Qurra (826-901), ce qui a particulièrement attiré l'attention de la philosophie arabe c'est le concept grec de « science harmonique » (Greek Ἀρμονική). Ceci est particulièrement présent dans les textes d'al-Kindī, et c'est ce qui constitue *ʿIlm al-taʾlīf* (science de la composition).

Cette « science » est liée à la musique en tant que champ théorique, mais sa source est la conscience de l'effet des sons sur l'âme, c'est-à-dire sur un monde non matériel. Rappelons que le principe fondamental de cette théorie, que nous retrouvons chez Pythagore (VIe siècle av. J-C.) et dans l'école néo-platonicienne, est que l'harmonie sonore dans la musique est due à est une adéquation entre les nombres et les proportions numériques. Par extension, ce modèle musical nous renseigne sur l'harmonie de tout l'univers non perceptible et métaphysique. Les instruments de musique doivent nous permettre

2 Būlos, transl. Kazimi (1999). Une version préliminaire de ce texte se trouve sur cette page https://saramusik.org/61/texte/ (URL accessed 17 April 2021).

3 Pour une discussion sur l'origine des deux textes *Adāb al-falāsifa et* et *Nawādir al-falāsifa*, cfr. Zakeri (2006) et http://saramusik.org/65/texte/ (URL accessed 17 April 2021).

4 Wright (1966), 27-48, ainsi que http://saramusik.org/16/texte/ (URL accessed 17 April 2021).

5 http://saramusik.org/60/texte/ (URL accessed 17 April 2021).

de façonner cette compatibilité. C'est ainsi qu'al-Kindī résume cela dans son *Kitāb al-Muṣawwiṭāt al-Watariyya* :

> وَكَانَتْ عَادَةُ الفَلَاسِفَةِ إِظْهَارُ أَسْرَارِ عِلْمِ الطَّبِيعَةِ وَآثَارِهَا فِي كَثِيرٍ مِنْ مَوْضُوعَاتِهِمْ فِي الكُتُبِ، مِنْ ذَلِكَ مَا سَمَّوْهُ فِي الأَرِتْمَاطِيقِي وَالأَعْدَادِ المُتَحَابَّةِ وَالمُتَبَاغِضَةِ، وَفِي الخُطُوطِ المُتَنَاسِبَةِ، وَفِي المُجَسَّمَاتِ الخَمْسَةِ الوَاقِعَةِ فِي الكُرَةِ. فَلَمَّا أَبَانُوا أَنَّهُ لَيْسَ شَيْءٌ مَحْسُوسًا إِلَّا وَعُنْصُرُهُ مُتَكَوِّنٌ مِنَ الأَرْكَانِ الأَرْبَعَةِ وَالطَّبِيعَةِ الخَامِسَةِ، أَعْنِي النَّارَ وَالهَوَاءَ وَالمَاءَ وَالأَرْضَ وَالفَلَكَ، بَعَثَتْهُمِ الفِطْنَةُ وَدَلَّهُمِ الذَّكَاءُ وَأَطْلَعَهُمِ الفِكْرُ عَلَى إِيدَاعِ آلَاتٍ صَوْتِيَّةٍ وَتَرِيَّةٍ، تُوَسِّطُ بَيْنَ النَّفْسِ وَبَيْنَ تَأْلِيفِ العَنَاصِرِ وَالطَّبِيعَةِ الخَامِسَةِ بِاتِّخَاذِ الآلَاتِ، وَصَنَعُوا آلَاتٍ وَتَرِيَّةً كَثِيرَةً تُنَاسِبُ تَأْلِيفَ الأَجْسَادِ الحَيَوَانِيَّةِ، وَيَظْهَرُ مِنْهَا أَصْوَاتٌ مُشَاكِلَةٌ لِلتَّرْكِيبِ الإِنْسِيِّ لِيُظْهِرُوا بِذَلِكَ لِلْعُقُولِ الذَّكِيَّةِ مِقْدَارَ شَرَفِ الحِكْمَةِ وَفَضْلِهَا[6]

> Les Philosophes avaient pour habitude de dévoiler les secrets de la science de la Nature et ses effets dans plusieurs livres. Ceci est par ce qu'ils appellent Arithmétique et nombres amiables et non amiables, ainsi que par les cinq corps solides qu'on peut placer dans une sphère. Et lorsqu'ils ont montré que la composition de tout corps sensible est constituée des quatre éléments plus le cinquième, je veux dire le feu, l'air, l'eau, la terre et l'éther, ils ont eu l'intelligence et le discernement pour développer des cordophones qui forment l'intermédiaire entre l'esprit et la composition des éléments, plus le cinquième élément. Ainsi, ils ont fabriqués plusieurs instruments à cordes similaires à la composition des corps animaliers ; ils produisent des sons conformes à la composition humaine ce qui montre aux esprits vifs l'importance de la sagesse et son utilité.

Puisqu'il est possible grâce à l'arithmétique de calculer les relations entre les sons musicaux et d'étudier leur propriétés, il serait alors possible de créer un effet sur l'auditeur et de choisir les différentes compositions. À noter que le concept de *ta'līf* qui correspond dans ce contexte à « harmonie », sera réduit la plupart du temps à l'idée de « composition », et notamment « composition des mélodies par agencement des notes »[7].

6 http://saramusik.org/6/texte/ (URL accesed 17 April 2021). Concernant le néo-platonisme d'al-Kindī, cfr. Adamson (2007).

7 تأليف الألحان بتركيب النّغمات

C'est donc à partir de ces principes que les premiers textes arabes sur la musique se sont élaborés, notamment ceux qu'al-Kindī, qui, avec l'*Épître sur la musique*, partage le même esprit néo-platonicien des Ikhwān al-Ṣafā'[8].

Al-Kindī fait référence également au *'Ilm al-ta'līf* dans son épître sur « la quantité des livres d'Aristote et sur ce dont on a besoin pour acquérir la Philosophie »[9], le plaçant aux côtés de l'arithmétique, comme étant la deuxième science qui traite de la quantité, contrairement aux sciences de la forme (*al-kayfiyya*), qui sont la géométrie et l'astronomie[10]. Suivant les traditions antérieures, il donne également la première place à l'arithmétique, la deuxième à la géométrie, la troisième à l'astronomie/astrologie, considérée comme un composée d'arithmétique et de géométrie, et la quatrième place à l'harmonique, considérant qu'elle est présente dans toutes les sciences mathématiques.

فَإِنَّ التَّأْلِيفَ فِي الكُلِّ، وَأَظْهَرُ مَا يَكُونُ فِي الأَصْوَاتِ وَتَرْكِيبِ الكُلِّ وَالأَنْفُسِ الإِنْسَانِيَّةِ، كَمَا أَثْبَتْنَا فِي كِتَابِنَا الأَعْظَمِ فِي التَّأْلِيفِ[11]

> L'harmonique est partout ; mais elle apparaît principalement à travers les sons, et la composition de tout avec l'âme humaine, comme nous l'avions montré dans notre *Plus grand livre de l'harmonique*.

Ainsi il apparaît clairement que dans cette vision la science harmonique est une science générale et le l'univers sonore est l'une de ses formes.

2 La musique comme science mathématique

Quant à l'idée que la musique est une science mathématique, cela était parfaitement intégrée au IXe siècle suite à la lecture des textes grecs, puisque la musique dans ce cadre constituait l'aspect numérique de la science harmonique. Dans son *Kitāb al-muṣawwiṭāt al-watariyya*, Al-Kindī nous rappelle le statut intermédiaire des sciences mathématiques :

---

8 Wright (2010).

9 Abu Rīda (1950), 359-384.

10 Abu Rīda, Ibid., 377.

11 Abu Rīda, Ibid. (1950). Cet ouvrage de al-Kindī [*Rasā'il al-Kindī al-falsafiyya*] cité par Ibn Nadīm (Xe siècle) dans son *Fihrist*, est aujourd'hui perdu.

Les Philosophes avaient pour habitude de se focaliser sur la science médiane, avec une science qui lui supérieure et une autre inférieure. La science qui lui est inférieur est la science de la Nature et ce qui en dérive, alors que celle d'en dessus est la Métaphysique, dont on voit la trace dans le monde physique. Quant à la science médiane, qui dérive vers les savoirs supérieur et inférieur, elle se compose de quatre classes : 1. La science du nombre et du dénombrable ; c'est l'Arithmétique ; 2. La science de l'Harmonique, qui est la Musique ; 3. La Géométrie, qui est la *Handasa* ; 4. La science de l'Astronomie, qui est le *Tanjīm*.

إِنّ عَادَةَ الفَلَاسِفَةِ كَانَتْ الارْتِيَاضُ بِالعِلْمِ الأَوْسَطِ، بَيْنَ عِلْمٍ تَحْتَهُ وَعِلْمٍ فَوْقَهُ. فَأَمّا الّذِي تَحْتَهُ فَعِلْمُ الطّبِيعَةِ وَمَا يَنْطَبِعُ عَنْهَا، وَأَمّا الّذِي فَوْقَهُ فَيُسَمّى عِلْمَ مَا لَيْسَ مِنَ الطّبِيعَةِ، وَتَرَى أَثَرَهُ فِي الطّبِيعَةِ. وَالعِلْمُ الأَوْسَطُ، الّذِي يَتَسَبّلُ إِلَى عِلْمِ مَا فَوْقَهُ وَمَا تَحْتَهُ يَنْقَسِمُ إِلَى أَرْبَعَةِ أَقْسَامٍ وَهِيَ : 1. عِلْمُ العَدَدِ وَالمَعْدُودَاتِ وَهْوَ الأَرِتْمَاطِيقِي ؛ 2. عِلْمُ التّأْلِيفِ وَهْوَ المُوسِيقَى ؛ 3. عِلْمُ الجَاوِمطْرِيّةِ وَهْوَ الهَنْدَسَة ؛ 4. عِلْمُ الأَسْطرُنُومِيّةِ وَهْوَ التّنْجِيم.

Les principes mathématiques forment donc le fondement de l'Harmonique. Suivant la tradition pythagoricienne, ils sont étroitement liés aux nombres 1, 2, 3 et 4, qui forment les premières proportions musicales : les nombres 1 et 2 forment l'intervalle du Tout, 2 et 3 l'intervalle du même et un tier, appelé dans le domaine musical quinte, 3 et 4 l'intervalle du même et un quart, appelé quarte. Puisque les nombres sont infinis, le premier exercice consistait donc à retrouver ces proportions entre différents nombres et puis à étudier la nature des différentes relations entre les nombres naturels.

Ces principes étaient connus de Pythagore et ses successeurs, les pythagoriciens, qui l'ont développé dans un contexte mathématique, indépendant des sons. Cet éloignement de l'aspect pratique, au point d'aboutir à du non-sens, leur a valu des critiques sévères de la part d'Aristoxène de Tarente (IVe siècle av. J-C.) élève d'Aristote[12]. Cependant, les mathématiciens d'Alexandrie, Euclide (fl. 300 av. J-C.) et Ptolémée (fl. IIe siècle), ont révisé par la suite cette théorie musicale, démontrant la correspondance entre les calculs et le sens auditif[13].

Il est important de noter que même si ces principes mathématiques étaient connus d'al-Kindī et d'Ikhwān al-Ṣafā nous ne retrouvons les éléments de continuité avec Euclid et Ptolémée dans les textes arabes qu'à la fin du dixième, avec le *Kitāb al-Mūsīqī al-Kabīr* d'al-Fārābī (878–950). L'école qui en dérive a écarté l'aspect astronomique et métaphysique de Ptolémée, en se focalisant sur la question de l'adéquation entre les mathématiques et le perceptible en musique remis en cause par l'école d'Aristoxène. Il en est de même pour Ibn

12 Cfr. Bélis (1986).

13 Les textes d'Euclide et de Ptolémée sont traduits dans Barker (1989). Voir également les éditions et traductions de Barbera (1991) et Solomon (2000).

Sīnā, dont l'épître sur la musique, *Jawāmiʿ ʿIlm al-Mūsīqā*, fait partie de la section des mathématiques dans le *Kitāb al-Shifāʾ*.

La découverte d'une version arabe des *Sectio Canonis* d'Euclide[14] est particulièrement importante, car ce texte démonstratif montre dans les premières propositions le passage du qualitatif au quantitatif dans le domaine sonore, et traduit explicitement ce qui constitue le fondement des mathématiques musicales :

Si nous sommes en repos et qu'il n'y a pas de mouvement, nous serons toujours en repos ;

- Si nous sommes en repos et que rien ne bouge, nous n'entendrons aucun son ;
- Si nous entendons un son, c'est que celui-ci est précédé d'un choc et d'un mouvement ;
- Et puisque toutes les notes ne peuvent exister que s'il y a eu choc, lui même issu d'un mouvement, et que le mouvement peut être rapide ou lent – les mouvements rapides produisent des notes plus aiguës et les mouvements lents produisent des notes plus graves –, par conséquent, les notes les plus aiguës sont engendrées par plus de mouvement et un nombre plus grand, alors que les notes les plus graves sont engendrées par moins de mouvement et un nombre plus petit[15] ;
- Les notes les plus aigües deviennent plus équilibrées par moins de mouvement, et les plus détendues le deviennent par plus de mouvement ;
- Alors, il est nécessaire de considérer la note comme étant composée de sections plus ou moins importantes selon le mouvement ;
- Et la relation entre tous les sons composés de sections traduit la relation entre leur nombre, on

- إِنْ كَانَ سُكُونٌ وَلَا حَرَكَةٌ كَانَ سُكُونٌ ؛
- وَإِذَا كَانَ سُكُونٌ وَلَمْ يَتَحَرَّكْ شَيْءٌ لَمْ يُسْمَعْ شَيْءٌ ؛
- وَإِنْ سُمِعَ شَيْءٌ وَجَبَ أَنْ يَكُونَ قَبْلَهُ قَرْعٌ وَحَرَكَةٌ ؛
- وَلِأَنَّ جَمِيعَ النَّغَمِ يَكُونُ إِذَا قُرِعَ وَلَا يُمْكِنُ أَنْ يَكُونَ إِلَّا وَقَبْلَهُ حَرَكَةٌ، وَالحَرَكَةُ مِنْهَا أَكْثَرَ حَثًّا وَأَقَلَّ حَثًّا وَالَّتِي هِيَ أَكْثَرُ حَثًّا تَفْعَلُ نَغَمًا أَكْثَرَ حِدَّةً، وَالَّتِي هِيَ أَقَلُّ حَثًّا تَفْعَلُ نَغَمًا أَكْثَرَ ثِقَلًا فَيَجِبُ بِاضْطِرَارٍ أَنْ يَكُونَ النَّغَمُ أَكْثَرَ حِدَّةً لِأَنَّهَا مِنْ حَرَكَةٍ أَكْثَرَ حَثًّا وَأَكْثَرَ عَدَدًا وَأَنْ يَكُونَ النَّغَمُ أَكْثَرَ ثِقَلًا لِأَنَّهَا مِنْ حَرَكَةٍ أَقَلَّ حَثًّا وَأَقَلَّ عَدَدًا ؛
[...]
- فَالنَّغَمُ الَّتِي هِيَ أَكْثَرُ حِدَّةً تَصِيرُ إِلَى الاعْتِدَالِ بِنُقْصَانِ الحَرَكَةِ وَالَّتِي هِيَ أَكْثَرَ اسْتِرْخَاءً تَصِيرُ إِلَى الاعْتِدَالِ بِزِيَادَةِ الحَرَكَةِ ؛
- فَلِذَلِكَ يَنْبَغِي أَنْ يُقَالَ إِنَّ النَّغَمَ مُؤَلَّفٌ مِنْ أَجْزَاءَ إِذَا كَانَتْ تَصِيرُ إِلَى الاعْتِدَالِ بِالزِّيَادَةِ وَالنُّقْصَانِ فِي الحَرَكَةِ ؛
- وَجَمِيعُ الأَصْوَاتِ الَّتِي هِيَ مُؤَلَّفَةٌ مِنْ أَجْزَاءَ فَإِنَّ نِسْبَةَ بَعْضِهَا إِلَى بَعْضٍ نِسْبَةُ عَدَدٍ إِلَى عَدَدٍ فَالنَّغَمُ إِذًا بِاضْطِرَارٍ يُقَالُ بَعْضُهَا إِلَى بَعْضٍ بِنِسْبَةِ عَدَدٍ إِلَى عَدَدٍ ؛

14 Neubauer (2005).

15 Il est clair ici que la quantité de mouvement traduit la fréquence et non pas l'amplitude.

doit donc considérer les notes selon les proportions qui existent entre les nombres[16] ;

– Les nombres peuvent être en relation de multiples, du même et une partie, ou des mêmes et plusieurs parties. Les notes de musique doivent donc être décrites par des proportions similaires ; celles ayant les mêmes proportions doivent être dénommées de la même manière[17] ; [...]

– وَالأَعْدَادُ مِنْهَا مَا يُقَالُ بِنِسْبَةِ أَضْعَافٍ وَمِنْهَا مَا يُقَالُ بِنِسْبَةِ مِثْلٍ وَجُزْءٍ وَمِنْهَا مَا يُقَالُ بِنِسْبَةِ مِثْلٍ وَأَجْزَاءٍ. فَالنَّغَمُ إِذًا بِاضْطِرَارٍ يُقَالُ بَعْضُهَا إِلَى بَعْضٍ بِهَذِهِ النِّسَبِ وَمَا كَانَ مِنْ هَذِهِ بِأَضْعَافٍ أُولَائِكَ وَحَرَكَاتِهِ يُقَالُ بَعْضُهَا إِلَى بَعْضٍ بِاسْمٍ وَاحِدٍ، وَأَمَّا مَا كَانَ بِمِثْلٍ وَجُزْءٍ فَلَيْسَ كَذَلِكَ ؛ [...]

Les principes fondamentaux des mathématiques musicales consistent donc en l'élaboration des règles numériques pour le calcul des intervalles musicaux, autrement-dit les proportions, et la présentation des méthodes qui permettent de :

– Additionner et soustraire les intervalles ;
– Multiplier et diviser les intervalles[18].

En adoptant, par convention, le concept de (*jins*), intervalle de quarte (4/3) contenant trois intervalles, il devient possible d'étudier les différentes combinaisons d'intervalles qu'ils peut contenir, développant ainsi la théorie de genres doux et forts[19].

3 Musique et Science de la Nature

Comme évoqué plus haut, Ibn Sīnā définit la musique comme une science mathématique. Selon lui celle-ci est constituée de deux objets d'étude (*mabāḥith*) : 1. *'Ilm al-Ta'līf*, ici au sens strict de « composition des mélodies » ; 2. *'Ilm al-īqā'*, la science du Rythme. Chacun de ces domaines possède des principes issus d'autres sciences, qu'ils soient arithmétiques, naturels (autrement-dit relatifs à la science de la Nature, la physique) ou géométriques. Cependant, Ibn Sīnā attire l'attention particulièrement sur les principes naturels, puisque l'objet de la musique, le son, appartient à la Nature. Ainsi, « s'il faut prendre des conclusions concernant cette science, en partant de postulats, ceux-ci ne peuvent être qu'issu de la Physique. Les principes numériques ne concernent

16 Euclide décrit ici ce qu'on traduit dans le vocabulaire musical par « intervalle ».

17 L'octave (2/1), la quinte (3/2) et la quarte (4/3) par exemple, doivent garder les mêmes noms, même si elles sont représentées par les relations 12/6, 12/8, 12/9, etc.

18 Pour plus d'informations concernant ces règles de calcul, cfr. Ghrab (2009), 56.

19 Concernant la formation des genres chez al-Fārābī, Ibn Sīnā et al-Urmāwī, Ibid., 61-68.

cette science que par la forme que prend son objet ».[20] Cette orientation épistémologique aura un impact sur l'un des plus importants théoriciens italiens au XVIe siècle, Gioseffo Zarlino (1517-1590), qui cite Ibn Sīnā sur ce même sujet[21].

Par ailleurs, cette tendance à observer les propriétés physiques du son est manifeste suite à la discussion sur le son qui s'était déclenchée à travers les commentaires au *Kitāb al-Adwār*[22] de Ṣafī al-Dīn al-Urmawī (1216-1294). Celle-ci trouve son origine dans le texte d'al-Fārābī, thématique issu de la tradition aristotélicienne du *Traité de l'âme*. Le problème en question était lié à la relation qu'entretiennent les corps mis en jeu pour la production du son. Le son était-il rattaché au corps qui frappe ou au corps frappé ? Dans sa réponse, l'Anonyme 61 a critiqué al-Urmawī dans sa position par rapport à al-Fārābī[23], ce qui l'a amené à résumer certains éléments liés au son, à savoir, sa constitution en forme de vagues suite à un choc, qui aboutissent à l'oreille, « ce qui provoque une réaction du nerf touché par le tympan auditif placé sur le conduit auditif[24] ». Le même sujet est abordé dans le commentaire de l'Anonyme 62[25] :

وَلَمَّا كَانَ الصَّوْتُ جِنْسًا لِلنَّغْمَةِ احْتَجْنَا إِلَى تَعْرِيفِهِ
وَهُوَ كَيْفِيَّةٌ تُحَسُّ بِهَا السَّامِعَةُ عِنْدَ اصْطِكَاكِ جِسْمَيْنِ،
وَالاصْطِكَاكُ مُصَادَمَةٌ مَعَ مُقَاوَمَةِ الْمَزْحُومِ لِلزَّاحِمِ.
وَأَمَّا السَّمْعُ فَلَا يُحْتَاجُ إِلَى تَعْرِيفٍ لِأَنَّ مَنْ لَهُ هَذِهِ
الْقُوَّةَ فَعِنْدَهُ بَدِيهِيٌّ وَمَنْ لَيْسَتْ لَهُ فَلَا يَتَصَوَّرُهُ
بِالتَّعْرِيفِ. وَأَمَّا كَيْفِيَّةُ إِحْسَاسِ السَّامِعَةِ بِالصَّوْتِ
فَإِنَّ الْهَوَاءَ إِذَا انْضَغَطَ وَنَبَا مِنْ بَيْنِ الْجِسْمَيْنِ الْمُصْطَكَّيْنِ
يُحَرِّكُ الْجُزْءُ الْمُجَاوِرُ لَهُ بِسَبَبِ نُبُوِّهِ فَيَقْبَلُ الْكَيْفِيَّةَ الَّتِي
قَبِلَهَا الْجُزْءُ الْأَوَّلُ مِنَ الْمُصَادَمَةِ وَيُحَرِّكُ الثَّانِي الثَّالِثَ
فَيَقْبَلُ مَا قَبْلَهُ وَالثَّالِثُ رَابِعًا، فَلَا يَزَالُ هَذَا التَّنَاوُلُ مِنْ
جُزْءٍ إِلَى جُزْءٍ مِنْ جَمِيعِ جَوَانِبِ مَوْضِعِ الْمُصَاكَّةِ، إِنْ

Le son étant le genre de la note, il nous faut ainsi le définir : Le son est une forme que notre sens de l'ouïe perçoit lorsque deux corps se heurtent. Le heurt est une rencontre dans laquelle le corps qui reçoit la poussée résiste à celui qui l'imprime. Quant à l'ouïe, nous n'avons pas besoin d'en donner la définition. Celui qui possède ce sens conçoit nécessairement ; et celui qui en serait dépourvu ne saurait se le figurer à l'aide d'une définition. Voici maintenant comment se poursuit la perception du son : Lorsque l'air qui se trouve entre les deux corps qui se heurtent est comprimé, il rebondit et meut la couche d'air immédiatement à sa suite. Cette dernière meut la couche suivante

20 Ibn Sīnā, « *Jawāmiʿ ʿIlm al-Mūsīqā* » (1956), 9-10. Il nous dit également à la septième section du *Kitāb al-Burhān* : « L'objet de la musique est un attribut de genre pour l'objet de la physique », Ibn Sīnā, *Kitāb al-Burhān* (1965), 162, voir également 164-165.

21 Corwin (2008), 323-324.

22 Ghrab (2009).

23 Ibid., 27-38, ainsi que les pages 247-251 pour le texte arabe.

24 Ghrab (2009), 250 : « فَيَنْفَعِلُ العَصَبُ الذي يدعن بِطبلةِ السامعةِ المفروشةِ على الصِماخ ».

25 La personnalité de l'Anonyme 62 est encore non identifiée, même si est considéré dans la traduction de d'Erlanger (1938), Vol. 4, comme étant Mawlānā Mubārak Shāh, et al-Jurjānī dans certaines copies. Pour plus d'informations sur ce texte, cfr. Ghrab (2009), 11-12.

d'un mouvement pareil à que lui a imprimé le choc [initial]. LA deuxième couche en meut une troisième ; celle-ci reçoit le mouvement tel que l'a reçu la deuxième, puis le communique à une quatrième, et ainsi de suite. Ce mouvement va toujours en se propageant, se répandant autour du point où le choc s'est produit et dans toutes les directions. S'il ne rencontre pas d'obstacle dans une direction ou dans une autre, il finit par atteindre la couche d'air qui remplit le conduit auditif, soit celle en contact avec l'organe qui siège de la puissance auditive. C'est alors que cette dernière perçoit le son. Ce mouvement se fait en ondes circulaires, semblables à celles qui se produisent à la surface d'une eau stagnante lorsqu'on y jette une pierre.

Si aucun obstacle ne vient s'opposer au mouvement d'ondulation, tu pourras te figurer la forme de la masse d'air soumise à ce mouvement en considérant celle d'un œuf. La partie la plus épaisse de cette figure est une grande circonférence parallèle au cercle d'horizon du lieu où s'est produit le heurt. Au-dessous de cette circonférence s'en superposent plusieurs autres, parallèles entre elles et à la plus grande. Chacune d'elles étant plus petite que celle qui est au-dessous d'elle, tout comme les cercles de latitude dans la sphère, elles finissent en un point. Il en va de même au dessous de la plus grande circonférence, avec cette différence que les cercles supérieurs sont respectivement plus étendus que les cercles inférieurs. La droite qui joindrait le centre de la grande circonférence au point supérieur serait de ce fait plus grande que celle qui réunirait le centre de la grande circonférence au point inférieur. La nature de l'air est, en effet, telle qu'il s'oppose au mouvement contraint qui pousse à la descente, tandis qu'il facilite le mouvement ascendant.[26]

لَمْ يكُنْ مانِعٌ فِي جِهَةٍ، حَتَّى يَنْصَدِمَ الجزْءُ المَوْجُودُ فِي الصِّماخِ المُجاوِرِ للعضْوِ الّذِي فِيهِ القوّةُ السّامِعَةُ فتحِسّ بِهِ، كَما يُشاهَدُ فِي تَمَوُّجِ المَاءِ الرَّاكِدِ عَلَى اسْتِدَارَةٍ عِنْدَ إِلْقَاءِ الحَصَاةِ فِيهِ وَاحسِبْ شَكْلَ الهواءِ المُتَمَوّجِ عِنْدَ [انعِدامِ]عدمِ المانعِ شَكْلَ بَيْضٍ أَغْلَظُ أَجْزَائِهِ دائِرَةٌ عَظِيمَةٌ مُوازِيَةٌ لِدائِرَةِ أُفُقِ مَوْضِعِ المُصاكَّةِ، وفَوْقَها دَوائِرُ مُتَوازِيَةٌ ومُوازِيَةٌ للعَظِيمَةِ، كُلُّ واحِدَةٍ أَصْغَرُ مِمَّا تَحْتَها كالمُقَنْطَرَاتِ ⟨١٧٣أ⟩ إِلَى أَنْ تَنْتَهِي إِلَى نُقْطَةٍ وكَذَلِكَ فِيمَا تَحْتَ الدَّائِرَةِ العَظِيمَةِ إِلَّا أَنَّ الدَّوائِرَ الفَوْقَانِيَّةَ أَعْظَمُ مِنَ التَّحْتانِيَّةِ فَيكُونُ الخطُّ الواصِلُ بَيْنَ النُّقْطَةِ الفَوْقانِيَّةِ ومَرْكَزِ العَظِيمَةِ أَطْوَلَ مِنَ الواصِلِ بَيْنَ التَّحْتانِيَّةِ والمَرْكَزِ، وذَلِكَ لِمُعَاوَقَةِ طَبِيعَةِ الهَواءِ لِلْحَرَكَةِ القَسْرِيَّةِ الَّتِي إِلَى تَحْتُ ومُعَاوَنَتِها للَّتِي إِلَى فَوْقُ، ولَا يَقْدَحُ فِيما أَوْرَدْنا ما يُقالُ أَنَّ مِنَ الإِسْطِقِسّاتِ إِذَا حَصَلَ فِي مَكانِهِ لَا يَبْقَى لَهُ مَيْلٌ إِلَى جِهَةٍ لِأَنَّ المُرادَ مِنَ المَكانِ نِهايَتُهُ وفِيمَا دُونَها لَا يَظْهَرُ المَيْلُ لِحَيْلُولَةِ الإِسْطِقِسِّ الآخَرِ أَوْ سائِرِ أَجْزاءِ نَفْسِهِ بَيْنَهُ وبَيْنَ وَجْهِهِ ولَوْ فُرِضَ ارْتِفاعُ الحائِلِ لَوُجِدَ المَيْلُ، ولَا يُقالُ إِنَّ المَيْلَ يَحْدُثُ بِارْتِفاعِ المَانِعِ لِامْتِنَاعِ أَنْ يَكُونَ أَمْرٌ عَدَمِيٌّ عِلَّةً لِأَمْرٍ وُجُودِيٍّ، فَلْيُتَأَمَّلْ. ويَتَغَيَّرُ هَذَا الشَّكْلُ بِهُبُوبِ الرِّياحِ وغَيْرِها مِنَ المَوَانِعِ ولِكَوْنِ التَّمَوُّجِ حَرَكَةً ذِيْنِيَّةً كُلَّما كانَ الصَّوْتُ مِنْ مَكانٍ أَبْعَدَ كانَ أَبْطَأَ وُصُولًا إِلَى السَّمْعِ كَما قَدَّمْنا فِي مِثالِ الصَّكَّةِ عَلَى القُلَّةِ ومِنْ هَهُنا يُسْتَدَلُّ عَلَى أَنَّ الإِحْساسَ بِالأَصْواتِ مِنْ طَرِيقِ التَّمَوُّجِ والتَّداوُلِ المَذْكُورَيْنِ.[27]

---

26 D'Erlanger, (1938), Tome III, 205-206.

27 British Library, MS Or. 2361, f. 72v-73r.

Il faut souligner également le fait que la tradition liée au son s'est perpétuée dans l'Occident musulman à travers les commentaires au *Kitāb al-Nafs* par Ibn Rushd (1126-1198)[28] et par Ibn Bājā (ob. 1138)[29], lecteur du *Kitāb al-Mūsīqī al-Kabīr* d'al-Fārābī. Ce dernier décrit également le phénomène de la résonance d'une corde[30] :

Par ailleurs, l'air, même s'il est lancé à partir du corps frappant, il reçoit également de celui-ci un impact qui lui est spécifique, comme on le voit à partir des corps vibratoires. Cela s'affiche clairement sur les cordes du *ʿūd* lorsque nous faisons vibrer la corde *bamm*, lors de l'accordage : ce qui se trouve sur la corde *mathnā* [en posant l'index] se met à vibrer, mais pas ce qui se trouve sur la corde *zīr*, ni sur la corde *mathlath*. De même, si la corde *mathlath* se met à vibrer, la corde *zīr* ne se mettra pas à vibrer, mais si on pose le doigt sur son index, il se mettra à bouger. Ceci se présente également entre les [notes] égales en hauteur, car elles sont semblables. Le même phénomène est présent aussi lors de l'[intervalle] du tout, où les [notes] se ressemblent mais ne sont pas égales.	وَالهَوَاءُ، مَعَ أَنَّهُ يَنْدَفِعُ عَنِ القَارِعِ ⟨يقبل⟩ عَنِ القَارِعِ أَثَرًا خَاصًّا بِهِ، كَمَا يَظْهَرُ ذَلِكَ مِنَ الأَجْسَامِ المُهْتَزَّةِ. وَبَيِّنٌ أَثَرُ ذَلِكَ الحِسِّ فِي أَوْتَارِ العُودِ فَإِنَّا نَجِدُهُ مَتَى حَرَّكْنَا البَمَّ فِي تَسْوِيَةِ المُطْلَقِ تَحَرَّكَ عَلَى المَثْنَى [بوضع السَّبَّابة]، فَلَمْ يَتَحَرَّكُ مَا عَلَى الزِّيرِ وَلَا مَا عَلَى المَثْلَثِ. وَكَذَلِكَ إِذَا اهْتَزَّ المَثْلَثُ لَمْ يَهْتَزَّ الزِّيرُ وَإِنْ وَضَعْنَا الإِصْبَعَ عَلَى سَبَّابَةِ الزِّيرِ تَحَرَّكَ مَا عَلَيْهِ. وَكَذَلِكَ يَعْرِضُ فِي المُتَسَاوِيَةِ الطَّبَقَةِ لأَنَّهَا مُتَشَابِهَةٌ، وَكَذَلِكَ عَرَضَ الأَمْرُ بِعَيْنِهِ فِيمَا بِالكُلِّ الَّذِي بِالكُلِّ مُتَشَابِهٌ وَلَيْسَ مُتَسَاوِي.

4 La musique parmi les sciences

Suite à l'intégration de la musique en tant que science mathématique, avec des réflexions concernant sa relation avec la physique, la sphère intellectuelle et scientifique ne pouvait pas ne pas réagir à un texte de théorie musicale sans prendre en considération cet aspect. La réaction de l'Anonyme XLI dans son commentaire au *Kitāb al-Adwār* d'al-Urmawī en est la meilleure illustration. Celui-ci dès le début du commentaire, s'attaque à al-Urmawī dans sa démarche :

---

28 http://saramusik.org/295/texte/ (URL accessed 17 April 2021).

29 http://saramusik.org/106/texte/ (URL accessed 17 April 2021).

30 La première description de ce phénomène apparaît dans les textes latins chez Decartes (1596-1650), *Compedium musicae* (1668).

Il dit: «Je l'ai organisé[31]». Je dis: Avant d'entamer le fond du sujet, il est nécessaire de savoir que chaque science possède un objet (*mawḍūʿ*), des principes (*mabādiʾ*) et des problèmes (*masāʾil*).

L'Anonyme XLI entame alors un exposé sur les objets des sciences et leurs principes en se basant sur le *Kitāb al-Burhān* d'Ibn Sīnā[32]. Il y explique comment celles-ci peuvent s'imbriquer, avant de catégoriser les principes des sciences en représentations (*taṣawwurāt*) et en vérités auxquelles on assentit (*taṣdīqāt*). Il évoque ensuite le fait que la musique a des principes arithmétiques et des principes provenant de la science de la Nature, afin de finir son exposé en abordant les problèmes des sciences[33]. Cette même lecture épistémologique se poursuit avec le deuxième commentateur du *Kitāb al-Adwār*, l'Anonyme XLII[34].

Pourtant, al-Urmawī calligraphe et musicien de cour présentait dans cet ouvrage ses connaissances de théorie musicale sans prétention à une quelconque activité scientifique. Celui lui a valu d'être le texte arabe sur la musique le plus lu et recopié. Cependant, il est évident qu'il n'avait pas connaissance du texte d'al-Fārābī rédigé trois ans plus tôt et ne se plaçait pas dans la tradition mathématique précédente, mais dans une tradition musicale savante profondément orale. Par contre, le deuxième texte que rédige al-Urmawī, la *Risāla Sharafiyya*[35], prend en considération l'ouvrage d'al-Fārābī, présentant ainsi une tentative de mise à jour de la théorie exposée dans le *Kitāb al-Adwār*. Il est bien possible que l'Anonyme XLI soit contemporain d'al-Urmawī et que ses critiques l'aient amené à rédiger son deuxième ouvrage. De ce fait la *Risāla Sharafiyya* cherche à être un livre scientifique alors que *Kitāb al-Adwār* est un livre de théorie musicale sans rattachement à une tradition scientifique.

5 La musique comme science indépendante

À partir de la *Risāla Sharafiyya* et des deux commentaires du *Kitāb al-Adwār* réalisé par l'Anonyme XLI et l'Anonyme XLII, les textes théoriques sur la musique commencent à se présenter en tant que textes scientifiques indépendants. Cette indépendance ne consiste pas en la rédaction d'ouvrage théorique

31 Ṣafī al-Dīn indique ici qu'il a organisé son ouvrage en quinze chapitres.
32 Ghrab (2009), 141-143 et 243-245.
33 Cfr. Jolivet (1997).
34 D'Erlanger (1938), Tome III 192-194.
35 D'Erlanger, (1938), Tome III.

sur la musique (ce type de texte existe déjà, comme le *Kitāb al-Mūsīqī al-Kabīr*) ou par le fait de ne pas inclure la musique dans les ouvrages encyclopédiques, à la manière d'Ikhwān al-Ṣafā, ou Abū ʿAbdallah al-Khawārizmī (750-850 ca.) ou Ibn Sīnā, mais par la nouvelle lecture épistémologique de la musique en tant que science qui cherche à bien définir son objet, ses principes et ses problèmes. Nous avions vu comment cela a été entamé en usant d'*al-Burhān*. Par la suite, deux auteurs adoptent cette nouvelle approche : Fatḥ Allāh al-Shirwānī[36] (ob. 1453 ca.) et ʿAbd al-Ḥamīd al-Lādhiqī[37] (ob. 1495 ca.).

La lecture attentive de la première version de l'épître d'al-Shirwānī[38], montre que suite l'exposé des différentes définitions de la musique depuis l'origine grecque du terme « Musique », l'auteur se focalise sur la définition de la science de l'Harmonique et sur sa relation avec la musique, comme donnée par l'auteur de l'ouvrage *al-Akhlāq al-Nāṣiriyya*[39]. Celui-ci considère que l'« étude des proportions harmoniques et de leurs propriétés » constitue l'Harmonique, et que si cela est appliquée aux sons « en considérant leur concordance et la quantité des silences qui les séparent », on l'appelle Musique. Ainsi, cet auteur considère que l'« objet de l'Harmonique est ce qui se présente aux proportions harmoniques, que cela soit sonore ou pas »[40]. Ensuite, même en adoptant une classification néo-platonicienne des sciences, il intègre deux aspects nouveaux :

1. La présentation des irrationnels parmi les nombres. Cela est en effet nécessaire dans la réflexion aristoxénienne liée à l'égalité des intervalles musicaux, impossible à réaliser dans le cadre des entiers naturels[41] ;
2. La discussion sur les propriétés physiques du son.

Par la suite, nous retrouvons le schéma présenté par al-Urmawī dans la *Risāla Sharafiyya*, et plus particulièrement dans la version étendue de ce texte[42].

L'approche d'al-Lādhiqī est similaire. Même s'il considère la musique comme une science mathématique, sa présentation systématise l'exposé de cette science, avec un objet, des principes et des problèmes relevant de deux

36 Personnage faisant partie de la migration intellectuelle de Samarkand vers l'Anatolie. Cfr. Neubauer, *Codex on Music* ainsi que http://saramusik.org/29/texte/ (URL accessed 17 April 2021).

37 Al-Lādhiqī, *Al-Risāla al-Fatḥiyya*, ed. al-Rajab (1986). Voir également http://saramusik.org/324/texte/ (URL accessed 17 April 2021).

38 Il y a deux versions de cette épître, cfr. Ghrab (2009), 12-15.

39 Nous ne sommes pas arrivés à déterminer l'auteur de cet ouvrage qui nous est inconnu.

40 Al-Shirwānī, ed. Neubauer (1986), 17.

41 Cfr. Ghrab, *Commentaire anonyme*, 56-57.

42 Pour une comparaison entre ces textes, cfr. Ghrab, *Commentaire anonyme*, 12-15.

domaines : la science harmonique, ici limitée à l'étude des notes et du système mélodique, et la « science » du rythme ; les deux ayant des principes appartenant à d'autres sciences[43]. Il catégorise ensuite les sciences selon leur rapport à la Matière : les sciences divines n'ont pas besoin de Matière, la physique a besoin de la Matière dans le monde réel et pour ses objets d'étude, alors que les sciences mathématiques ont besoin que la Matière existe, mais pas pour leurs objets d'étude. Cette classification lui permet d'entamer la présentation de la science de la Musique en ayant des principes physiques[44], qu'il place en premier, ensuite des principes arithmétiques et géométriques[45][46].

6 Conclusion

À partir du XIIe s., quelques traités de musiciens présentant des théories liées de manière plus directe à la pratique musicale commencent à voir le jour. Les traités et ouvrages que nous avons abordés tout au long de cet article ne représentent donc pas toutes les catégories d'ouvrages sur la musique rédigés en langue arabe durant la période s'étalant du XIe au XVe s. Ceux-ci se rattachent à une tradition de textes théoriques, rédigés dans la sphère intellectuelle et scientifique de leur époque, et leur approche évolue selon les différentes visions du monde.

Nous avons vu comment la première école représentée par al-Kindī incarne le néoplatonisme, où la Musique constitue le lien entre l'univers métaphysique et le monde physique : elle constitue dans ce cadre une science de l'Harmonique. En second lieu, c'est le renouveau du pragmatisme mathématique dans la musique qui commence à apparaître avec al-Fārābī. Il avait au départ pour objectif de relier les mathématiques musicales au monde sonore et sensible, en reprenant les positions euclidienne et ptoléméenne. Cette position se pour-

43 Al-Lādhiqī, *Al-Risāla al-Fatḥiyya*, ed. Al-Rajab (1986), 37.

44 Al-Lādhiqī, Ibid., 42-45.

45 Al-Lādhiqī, Ibid., 46-58.

46 Soulignons que le vocabulaire utilisé et certaines phrases montrent une filiation directe entre les textes d'al-Shirwānī et d'al-Lādhiqī. Mise à part leur structure générale, les deux textes précisent une *ghāya* (« objectif ») et un *sharaf* (« noblesse ») pour la Science de la Musique dont ils donnent les mêmes exemples : 1. La solidité des arguments liés à la musique car ils utilisent les preuves issus des mathématiques et de la physique ; 2. Évocation d'un propos attribué au prophète Mohamed concernant l'utilité de la Musique pour améliorer la voix et la psalmodie coranique ; 3. L'exemple d'un musicien qui a transformé par son art une querelle entre musiciens en une paisible amitié (Al-Shirwānī 20-21 ; Al-Lādhiqī, 40-41).

suit avec Ibn Sīnā, qui en donne une forme concise et précise, ce qui lui permet d'être facilement reproduite par les auteurs qui lui ont succédés. Parallèlement aux règles musico-mathématiques, Ibn Sīnā présente un cadre épistémologique qui sera utilisé par les Anonymes LXI et LXII afin de relire le texte de théorie musicale pratique le plus en vogue à partir du XIIIe s., le *Kitāb al-Adwār* d'al-Urmawī.

La tradition aristotélicienne liée au son, reprise notamment par al-Fārābī, formera sans doute l'aspect le plus marquant dans les textes du XIVe et XVe s. En utilisant le cadre épistémologique d'Ibn Sīnā, ces textes intègrent entièrement des principes physiques comme éléments fondamentaux de la Science de la Musique. Même si les traités d'al-Shirwānī et d'al-Ladhiqī n'auront pas beaucoup d'impact par la suite, car les textes arabes qui verront le jour à partir de cette période-là seront principalement des textes de théorie musicale indépendante du contexte scientifique, ils représentent les jalons d'une nouvelle tradition de textes indépendants sur la Science de la Musique. Cette nouvelle vision évoluera principalement en Europe, notamment avec les *Le Istitutioni Harmoniche* de Zarlino, l'*Harmonices Mundi* de Kepler, et les ouvrages de Marin Mersenne et de Descartes. Nous espérons donc avoir donnés ici des éléments utiles à une lecture plus globale et interculturelle de cette histoire de la Science de la Musique.

Bibliography

Abu Rida, Mohamed Abu al-Hadi, *Rasāʾil al-Kindī al-Falsafiyya* (Le Caire: Dār al-Fikr al-ʿArabī, 1950).

Adamson, Peter, *Al-Kindi* (New York: Oxford University Press, 2007).

Anonyme LXII, *Commentaire du Kitāb al-adwār* (MS. Or. 2361, London: British Museum), f. 68b-153a, cfr. 69a-70a; transl. d'Erlanger, Baron Rodolphe, *La Musique Arabe*, Tome III: *Ṣafiyyu-D-Dīn al-Urmawī: I. Aš-šarafiyya ou Epitre à Šarafu-d-dīn. II. Kitāb al-Adwār ou Livre des Cycles musicaux* / Traduction française, Vols. I-VI, Vol. III (Paris: Librairie Orientaliste Paul Geuthner, 1938).

Barbera, André, *The Euclidean division of the Canon: Greek and Latin Sources* (University of Nebraska Press, 1991).

Barker, Andrew, *Greek Musical Writings II*: Harmonic and Acoustic Theory, (Cambridge: Cambrige University Press, 1989).

Bélis, Annie, *Aristoxène de Tarente et Aristote: Le Traité d'Harmonique* (Paris: Klincksieck, 1986).

Chabrier, Jean-Claude. «Science musicale», dans *Histoire des sciences arabes*, Tome II (Paris: Seuil, 1997), 231-262.

Corwin, Lucille, « Le Istitutioni Harmoniche of Gioseffo Zarlino », Part 1 : A Translation with Introduction, PhD dissertation (New York : City University of New York, 2008).

D'Erlanger, Rodolphe, *La Musique Arabe*, Tomes III et IV (Paris : Paul Geuthner, 1938-1939).

Ghrab, Anas, « Commentaire anonyme du Kitāb al-adwār : édition critique, traduction et présentation des lectures arabes de l' œuvre de Ṣafī al-Dīn al-Urmawī », PhD dissertation (Paris : Université Paris 4 – Sorbonne, 2009).

Ibn Sīnā, Abū ʿAlī, « *Jawāmiʿ ʿIlm al-Mūsīqā* [Compendium sur la science de la musique] », in *Kitāb al-Shifāʾ* [Livre de la guérison], ed. Zakariyya Yūsuf, Tome 1, révisé par A.F. Ahwānī et M.A. al-Ḥifnī (Le Caire : Maktabat Miṣr, 1956).

Ibn Sīnā, Abū ʿAlī, « al-Manṭiq », chap. *al-Burhān*, in *Kitāb al-Shifāʾ*, ed. Abū al-ʿAlāʾ ʿAfīfī, Vol. IX (Le Caire : al-hay'a al-ʿāmma li-šuʿūn al-ṭibāʿa al-ʿāmiriyya, 1965).

Jolivet, Jean, « Classifications des sciences », dans *Histoire des sciences arabes*, Vol. III (Paris : Seuil, 1997), 264-266.

Kazemi, Elke, *Die Bewegte Seele – Das Spätantike Buch das Wesen der Musik* (*Kitāb ʿUnṣur al-Mūsīqī*) *von Paulos/Būlos in Arabischer Übersetzung vor dem Hintergrung der Griechischen Ethoslehre* (Frankfurt am Main : Institut für Geschichte der Arabisch-Islamischen Wissenschaften, 1999).

Al-Lādhiqī, Muḥammad ibn ʿAbd al-Ḥamīd, *Al-Risāla al-Fatḥiyya*, ed. Hāshim Muḥammad al-Rajab, al-silsila al-turāṭiyya 13 (Le Koweit : al-majlis al-waṭanī li-al-aqāfa wa al-funūn wa al-ādāb, 1986).

Neubauer, Eckhard. « Die Euklid zugeschriebene 'Teilung des Kanon' in Arabischer Übersetzung », *Zeitschrift für Geschichte der Arabisch-Islamischen Wissenschaften* 15 (2004-2005), 309-385.

Shiloah, Amnon. *The Theory of Music in Arabic Writings* (*c. 900-1900*) *: Descriptive Catalogue of Manuscripts in Libraries of Europe and the U.S.A.* (München : G. Henle Verlag, 1979).

Al-Shirwānī, Fatḥ Allāh, *Fatḥ Allāh al-Shirwānī : Majalla fi'l-Mūsīqī* (*Codex on Music*), ed. Eckhard Neubauer, Vol. XXVI, C – Facsimile Editions (Frankfurt am Main : Institut für Geschichte der Arabischen-Islamischen Wissenschaften an der Johann Wolfgang Goethe-Universität, 1986).

Solomon, Jon, *Ptolemy Harmonics*, translation and commentary (Leiden : Brill, 2000).

Zakeri, Moshen (ed.). *Persian Wisdom in Arabic Garb : ʿAlī b. ʿUbayda al-Rayḥānī (D. 219/834) and his Jawāhir al-kilam wa-farāʾid al-ḥikam* (Leiden : Brill, 2006).

Ute Pietruschka, « Gnomologia : Syriac and Arabic Traditions », in Houari Touati (ed.), *Encyclopedia of Mediterranean Humanism* (Spring 2014), http ://www.encyclopedie-humanisme.com/?Gnomologia (URL accessed 17 April 2021).

Wright, Owen (ed. And trasnsl.), *Epistles of the Brethren of Purity. On music. An Arabic critical edition and English translation of EPISTLE 5* (Oxford : Oxford University Press in association with The Institute of Ismaili Studies, 2010).

CHAPTER 5

Traditional and Modern Science in an Age of Transition: ʿAlī Muḥammad Iṣfahānī and the Logarithm of Numbers

Zeinab Karimian

1 Introduction

Recent research has highlighted the important role played in the transmission of modern science to non-Western lands by scholars who had a firm basis in traditional sciences before getting acquainted with modern methods.[1] This was the case for mathematics in some Islamic countries, where the existence of an age-old tradition of research in mathematical sciences created a favorable atmosphere for the adaptation of new mathematical concepts by some scholars already well versed in traditional mathematics.

ʿAlī Muḥammad Iṣfahānī (1215/1800–1293/1876),[2] was a notable scientific figure of nineteenth century Iran who spent the formative years of his scientific and professional life in Isfahan, before getting acquainted with European sciences at the modern institution, Dār al-Funūn in Tehran.

Several Iṣfahānī's scientific works are written in the style of traditional mathematical treatises like *Miftāḥ al-Ḥisāb* by Ghīyāth al-Dīn Jamshīd Kāshānī (d. 832/1429) or *ʿUyūn al-Ḥisāb* by Muḥammad Bāqir Yazdī (d. before 1659/1069), whereas some of his other works seem to have been written in the style of the pedagogic books of Dār al-Funūn. Some of the works ascribed to Iṣfahānī by historians are not available today and therefore the attribution of certain innovations to Iṣfahānī remains inconclusive.

In this paper, we try to show how Iṣfahānī, with only a second-hand knowledge of the new concept of logarithm, attempted to improve the rules of interpolation used in the calculation of a table of logarithms, so that the values found in the tables would be in agreement with the results of calculation. To do so, Iṣfahānī would guess the rule used by the author of the tables and

1 Rashed (1992).

2 In this article the dates are given in both Islamic lunar (AH) years and Christian (AD) year or Christian only.

 | DOI:10.1163/9789004513402_007

thereby make his method of interpolation more accurate. As a result, several Persian historical sources, such as *The Diffused Articles* by Abū al-Ḥasan Furūghī, and *The Algebra of Khayyām* by Ghulāmḥusiyn Muṣāḥib, credit Iṣfahānī with the discovery of the method of the extraction of the logarithm of numbers. Nevertheless, scholars are divided on the veracity of this attribution: the historian of mathematics, Abu al-Qāsim Qurbānī, believes it to be spurious, while a later author, ʿAbd al-Ḥusiyn Muṣ-ḥafī, concludes that there may be an element of truth to this attribution on the basis of some historical evidence.

In what follows, we will survey the life and works of this mathematician, briefly outline the previous scholarly debates, and present an analysis of the available evidence in order to delineate what can be reliably attributed to Iṣfahānī. Although claims attributing the "reinvention" of logarithm to Iṣfahānī are baseless, we hope to show that Iṣfahānī's contribution is still significant since it sheds light on the complex relation between traditional and modern sciences in a period of transition.

2 Iṣfahānī's Life and Works

ʿAlī Muḥammad ibn Muḥammad Ḥusiyn Iṣfahānī, praised by some of his contemporaries as "Ghīyāth al-Dīn Jamshīd Thānī", was born in 1215/1800 in Isfahan and died in 1293/1876 in Tehran. His life can be divided into two periods: the first period being when he lived in Isfahan before his immigration to Tehran and, the second after his immigration to Tehran in about 1268/1851. All we know about the earlier period of Iṣfahānī's life is that he began to study the sciences in his native city, achieved a high level of mathematical knowledge, wrote important treatises like *The Completion of the Spring-heads* [*of Arithmetic*] (*Takmilat al-ʿUyūn*), and had several students, including his children.

His eldest son, ʿAbd al-Wahhāb (1250/1834–1289/1872), was an astronomer and his second son, ʿAbd al-Ghaffār, known as Najm al-Dawla (1255/1839 or 1259/1843–1326/1908), was an exemplary scientific figure of the Qajar period. He was not only among the earliest students to graduate from Dār al-Funūn, but also became one of the principal mathematics teachers of that school at an early age and produced textbooks on several mathematical disciplines to be used as course books in Dār al-Funūn. He also served as the royal astronomer for several decades.

Perhaps the most outstanding among Iṣfahānī's disciples is Muḥammad ʿAlī Ḥusaynī Qāʾinī Bīrjandī Iṣfahānī (1224/1809–1311/1893), another mathematician of the nineteenth century, who in some of his treatises and commentaries mentions Iṣfahānī as his master.[3] Besides his own original works,[4] there exist several important Arabic mathematical treatises—such as the Arabic translation of the *Conics* of Apollonius (*Kitāb al-Makhrūṭāt*),[5] some recensions of mathematical works by Naşīr al-Dīn Ṭūsī,[6] *Tanqīḥ al-Manāẓir* by Kamāl al-Dīn Fārsī,[7] and *Miftāḥ al-Ḥisāb* by Kāshānī,[8] etc.—in which one can find marginal notes and valuable comments by Qā'inī. His notes show that he was well versed in traditional mathematics. In fact, he knew some European mathematics as well, since he mentions certain European achievements in his own works.[9] Several famous mathematicians of the time were trained under the supervision of Qā'inī, such as Mīrzā ʿAbdullāh Rīyāḍī (d. 1311/1893), ʿAbd al-Ḥusiyn ibn Muḥammad Mūsavī Dizfūlī and the sons of Farhād Mīrzā, the Qajar prince. The fact that a good part of the scientific life of these mathematicians was spent in Isfahan indicates that this city was still active in classical scientific learning at the time and that Iṣfahānī had succeeded in establishing his own school of mathematics.

ʿAlī Qulī Mīrzā Iʿtiḍād al-Salṭanah, a prince of the Qajar dynasty and the minister of science at the time (from 1276/1860 to 1298/1881), considered himself a disciple of Iṣfahānī. Iʿtiḍād al-Salṭanah had a passion for science and scientists, and is considered an eminent figure of the Qajar era who played an important role in the transmission of modern sciences to Iran. In his *Translation and Commentary of a Part of* The Chronology of Ancient Nations (*Tarjumi wa Sharḥ-i Bakhshī az* Āthār al-Bāqīya) by Bīrūnī (362/973–442/1050), Iʿtiḍād al-Salṭanah claims that all he has done in this book are due to the help of the great Master—"the Bīrūnī of the time"—ʿAlī Muḥammad Iṣfahānī.[10] Fur-

3 Cfr. MSS Malik 0601, fol. 62ʳ and Majlis 15537, fol. 116ʳ.

4 Like *A Survey of Sine and Tangent* (about how to derive and apply the sine and tangent of the numbers from the tables), *Mashāriq al-'Aḍwā, Nahāya al-Īḍāḥ fī Sharḥ-i Bāb al-Masāḥa min al-Miftāḥ* (a long commentary on the chapter on areas and volumes of *Miftāḥ al-Ḥisāb* by Kāshānī), etc.

5 MS Mashhad, Astān Quds 5619.

6 Beside his original works, the 13th century mathematician, Naṣīr al-Dīn al-Ṭūsī, made new Arabic recensions of some classic mathematical sciences known as the Middle Books. These were studied between Euclid's *Element* and Ptolemy's *Almagest*. Qā'inī's notes are found in the margins of Ṭūsī's *Recension of Archimedes'* On the Sphere and the Cylinder in MS Majlis 6411; *Recension of Archimedes'* Lemmas in MS Mashhad, Astān Quds 29374; *Recension of Menelaus'* Spherics in MS Majlis 824.

7 MS Majlis 168.

8 MS Majlis 15537.

9 Cfr. the treatise *A survey of Sine and Tangent*, MS Central Library of University of Tehran 462, fol. 1ʳ.

10 MS Malik 1471, fol. 6ʳ.

thermore, in an extant autograph anthology, Iʿtiḍād al-Salṭanah praises Iṣfahānī's great scientific dignity and expertise and calls himself a student of Iṣfahānī.[11]

Another figure who can be considered a student of Iṣfahānī is Mīrzā Jaʿfar Mushīr al-Dawla. Although no historical source mentions him as a student of Iṣfahānī, we can include him among Iṣfahānī's students based on the evidence given by an interesting treatise entitled *The Questions Mīrzā Jaʿfar Khān Mushīr al-Dawla Put toʿAlī Muḥammad Iṣfahānī with the Answers.*[12] Mushīr al-Dawla was one of the first Iranians who were sent to England in about 1230/1815 to study modern sciences and engineering. He was not only a distinguished political figure (being the minister of Foreign Affairs for a certain period) but also the first person in Iran to give a more or less technical account of some elements of the new (i.e., Newtonian) mechanics.[13] The treatise mentioned above is striking because Mushīr al-Dawla—a person knowledgeable in modern sciences—addresses some scientific questions to Iṣfahānī—a scholar who was rather known for his mastery of the traditional sciences. The questions discussed in the treatise are in the fields of geometry, arithmetic, and physics (including gravity), and the answers by Iṣfahānī sometimes refer to the propositions from Euclid's *Elements* or to the introduction of *The Elements of Astrology* (*al-Tafhīm*) by Bīrūnī.

We are not sure when exactly Iṣfahānī moved to Tehran. According to a well-known historical source, it was Iʿtiḍād al-Salṭanah who invited Iṣfahānī to Tehran because of his knowledge and expertise.[14] In the previously mentioned anthology of Iʿtiḍād al-Salṭanah, the autograph note in which the author considers himself as a disciple of Iṣfahānī bears the date 27th of Dhu al-Qaʿdah 1268/12th September 1852. Furthermore, at the end of a passage in an anthology written by Iṣfahānī himself, it is noted that he has written down this passage on 11th Shawwāl 1269/18th July 1853 in the village Farahzad (a neighborhood in Tehran) while the disease of cholera spread all over the region.[15] Therefore, we can conclude that Iṣfahānī immigrated to Tehran before 1269/1853, i.e., a little after the inauguration of the first modern Iranian pedagogical establishment,

---

11 MS Majlis 1453, 26.

12 MSS Majlis 1453/42 and Majlis 81 (the folios are not numbered).

13 Masoumi Hamedani (2001), 634.

14 Ṣanīʿ al-Dawla (1363/1984), 261.

15 هذا ما جرى عليه قلم العبد الجانى على محمد بن محمد حسين الاصفهانى في ليلة الأحد العاشر من شهر شوال المكرم ١٢٦٩ في قرية فرح زاد على سبيل الاستعجال مع تراكم الهموم والغموم حين استيلاء الوبا على بلاد الرى وقى الله عباده منها.

(MS Majlis 81, the folios are not numbered).

Dār al-Funūn, in Tehran and before the nomination of Iʿtiḍād al-Salṭanah as the minister of science.

The foundation of Dār al-Funūn was part of the reform plans of Mīrzā Taghī Khān Farāhānī, known as Amīr Kabīr (1222/1807–1268/1852), the first prime minister of Nāṣir al-Dīn Shāh, but it was officially inaugurated in 29th December 1851, i.e., thirteen days before the assassination of Amīr Kabīr by the order of Nāṣir al-Dīn Shāh. The first instructors of Dār al-Funūn were invited from Austria and Italy in order to teach military sciences, engineering, pharmacy, medicine, surgery and geology; and the students were for the most part chosen from among noble families. French was the official language, since there were several French translators in Iran, including the students who had been previously sent to France. During the first years of the foundation of Dār al-Funūn, each instructor had a personal interpreter to translate his courses. Gradually, Iranian teachers also began to teach at Dār al-Funūn.

According to some historical sources, in the early years of Dār al-Funūn, Iṣfahānī accompanied Iʿtiḍād al-Salṭanah as the examiner of the students and the inspector of scientific affairs in his visits of the school.[16]

Iṣfahānī's most significant mathematical work is *The Completion of the Spring-heads [of Arithmetic]* (*Takmilat al-ʿUyūn*) which is in fact a "completion" of *Spring-heads [of Arithmetic]* (*ʿUyūn al-Ḥisāb*) of Muḥammad Bāqir Yazdī, the famous mathematician of the eleventh/seventeenth century. We do not know when exactly Iṣfahānī wrote this treatise,[17] but we are almost sure that it was completed before his immigration to Tehran. In this treatise, Iṣfahānī classifies the algebraic equations—with degrees less than or equal to three—into twenty-five groups consisting of six binomial equations, twelve trinomials and seven quadrinomials. This treatise which was written in Arabic, the traditional scientific language, is directly related to the algebraic tradition of Khayyām (439/1048–526/1132) and Sharaf al-Dīn Ṭūsī (d. 610/1213).[18] According to Roshdi Rashed:

> From an epistemological point of view, the most interesting fact to be learned from this work is that this mathematician, who evidently knew only a little about the development of mathematics in the eighteenth century, could arrive, only on the basis of his twelfth-century predecessors, at some results which were similar to those obtained by mathematicians of

16 Ṣanīʿ al-Dawla (1367/1988), II, 1083.

17 According to Rashed this treatise has been written in 1239/ 1824, Rashed (1992), 294.

18 Rashed, Ibid., 394. Nacéra Bensaou has worked on the edition with a French translation and commentary of this treatise as a part of her PhD thesis in Paris-Diderot University.

the seventeenth and eighteenth centuries; and to do this only through an arithmetical study of polynomial functions and not by analytical considerations.[19]

Another treatise written by Iṣfahānī is entitled *The Division of Sphere by Means of Plane Surfaces* (*Taqsīm-i Kura bi Suṭūḥ-i Mustawīya*). In this Persian treatise, of which an autograph manuscript—dated 1274/1858—is extant, the author intends to show how to calculate, based on some elements of a segment of a sphere (such as altitude, diameter, and chord), the volume of the segment. This manuscript includes long tables copied by Iṣfahānī's eldest son, ʿAbd al-Wahhāb.[20]

Another work by Iṣfahānī, entitled *The Extraction of the Table of Logarithm of Sines*. The problems discussed in this treatise show the extent to which Iṣfahānī was acquainted with European mathematics.

It is reported that Iṣfahānī also wrote treatises on music, alchemy, the science of numbers, and logarithms,[21] most of which seem to be lost. Moreover, the French diplomat and philosopher, Comte de Gobineau (1816–1882), writes in his diaries that Mullā ʿAlī Muḥammad knew well the theory of music, although he played no musical instruments.[22]

3 Attribution of the Invention of the Logarithm to Iṣfahānī

3.1 *Historical Background*

The attribution of invention of the logarithm to Iṣfahānī has long been an object of controversy among Iranian historians. One of the earliest texts to attribute the invention of logarithm to Iṣfahānī is by a scholar of the late 19th and the early 20th century, Abu al-Ḥassan Furūghī. In this text,[23] after having praised the high rank of Iṣfahānī in traditional mathematics, Furūghī says that Iṣfahānī discovered, "thanks to his abundant knowledge and his keen intelligence, on the basis of the elements of mathematics inherited from Iranian and Muslim scientists 80 years ago", many algebraic rules and the "foundations of

19 Rashed, Ibid., 400; English translation is taken from Masoumi Hamedani (2001), 638.
20 Cfr. MS Majlis 2138.
21 Bāmdād (1378/1999), 482–488.
22 Joseph Arthur Gobineau, *Trois ans en Asie* (*de 1855 à 1858*) (1905), 442.
23 This text, called *Scattered Papers* (or *Diffused Articles*) (*Awrāq-i Mushawwash*/ *Maqālāt Mukhtalifa*) and independently published in 1330/1912, was first published as a part of the *Persian Calendar1328/1910*, by Najm al-Dawla [Pākdāman (1353/1974), 346].

the science of logarithm". Furūghī adds that unfortunately Iṣfahānī's discoveries did not propagate, since they coincided with the introduction of modern Western sciences.[24] The text of Furūghī implies that the discoveries of Iṣfahānī, including his "discovery" of logarithm, went back to about 1250/1830, i.e., when he still lived in Isfahan.

Another example of this attribution is found in a work by the famous mathematician and historian, Ghulāmḥusiyn Muṣāḥib (1328/1910–1399/1979). Muṣāḥib cites ʿAbd al-Ghaffār Najm al-Dawla according to whom his father, Iṣfahānī, could discover the benefits of logarithm by contemplating on a specific passage which is found in the treatise *Spring-heads of Arithmetic* (*ʿUyūn al-Ḥisāb*) by Muḥammad Bāqir Yazdī "who had been a contemporary of Shāh Ismāʿīl or Shāh Sulaymān, the Safavid kings".[25] This is the passage to which Najm al-Dawka refers:

> If we want to know the [value of the] axis (*sahm*) of a given arc, we multiply the sine of the half of that arc by itself, and divide the result by the sine of thirty degrees, i.e. the half of the half of the diameter. The result of the division is the axis of that arc. For example: we want to know the axis of fifty degrees. We take the sine of twenty-five degrees, it is 96259482. We multiply it by 2, it becomes 192518964. In fact, it is equal to the square of the sine of twenty-five degrees. We subtract from it the sine of thirty degrees, which is 9698700; there remain 95529264; which is equal to the result of the division of the square of the sine of twenty-five degrees by the sine of thirty degrees. Therefore, it is the axis of fifty degrees.[26]

To rewrite this phrase into modern notation, we consider *AB*, the given arc. *AG* is then the desired axis (Figure 5.1).

---

24 Furūghī, *Awrāq-i Mushawwash yā Maqālāt Mukhtalifa* (1330/1951), 38–39.

25 Muṣāḥib (1317/1938), 160–161; Najm al-Dawla, *Jadāwil-i Lugārītm-i Aʿdād-i Ṣiḥḥāḥ az 1 tā 1000* (1292 AH), 8.

As a matter of fact, the citation of Najm al-Dawla—quoted by Muṣāḥib—is extracted from his treatise *The Tables of the Logarithm of the Integers from 1 to 1000* (*Jadāwil-i Logārītm-i Aʿdād-i Ṣaḥīḥ az 1 tā 1000*), preserved in Malik National Library 11615, 8–9.

26 "وإذا أردنا أن نعلم سهم قوس معلومه، نضرب جيب نصف تلك القوس في نفسه ونقسم الحاصل على جيب ثلاثين درجة، أعني نصف نصف القطر. فالخارج من القسمة سهم تلك القوس. مثاله: أردنا أن نعرف سهم خمسين درجة. أخذنا جيب كه درجة، فكان 96259482. ضعفناه، فصار 192518964. فهو بمنزلة مربع جيب كه درجة. نقصنا منه جيب ل درجة وهو 96989700، بقي 95529264. وهو بمنزلة الخارج من قسمة مربع جيب كه على جيب ل درجة. فهو سهم ن درجة" [Najm al-Dawla, *Jadāwil-i Lugārītm-i Aʿdād-i Ṣiḥḥāḥ az 1 tā 1000* (1292) AH/1875), 8]

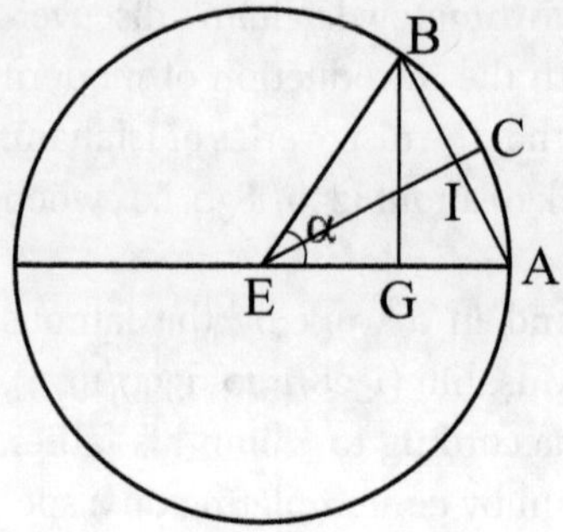

FIGURE 5.1

The triangles *ABG* and *AEI* are similar, so

$$\frac{AG}{AB} = \frac{AI}{AE}$$

$$\frac{AG}{\frac{1}{2}AB} = \frac{AI}{\frac{1}{2}AE} \Rightarrow AG = \frac{AI^2}{\frac{1}{2}r} = \frac{r^2 \sin^2 \frac{\alpha}{2}}{\frac{r}{2}}$$

for $\alpha = 50^\circ$ and $r = 1$:

$$\log(\sin 25^\circ) = -10 + 9.6259482$$

$$2\log(\sin 25^\circ) - \log(\sin 30^\circ) = \log\left(\frac{\sin^2 25^\circ}{\sin 30^\circ}\right)$$

$$\log AG = [(2 \times 9.6259482) - 9.6989700] - 10$$

$$\log AG = 9.5529264 - 10$$[27]

After having cited the passage in question, Najm al-Dawla remarks that when Mīrzā Naṣīr Ṭabīb—"a great scholar in many sciences who added marginal notes to this treatise of Yazdī"—fell upon this phrase, he wrote in the margin that he could not understand it. Najm al-Dawla continues that this remained unknown until his father discovered its concept and composed a treatise on how to extract the logarithm of numbers, and that this discovery was made in Isfahan in 1240/1824, that is, in a period when there were not yet any communications with Europeans.[28] It should be noted that Muṣāḥib adds some supplementary notes at the end of his book, one of which concerns the text of Najm al-Dawla on the specific passage of Muḥammad Bāqir Yazdī. Muṣāḥib

27 It should be noted that the numbers which are employed in the specific passage are in agreement with the numbers in the logarithm tables of Jean-Baptiste Morin [cfr. *Trigonometriae Canonicae Libri Tres* (1633), 168] or Henry Briggs' tables [cfr. Roegel, *A reconstruction of the tables of Briggs and Gellibrand's Trigonometria Britannica* (1633), (Research Report), 2010, 172) which are in the excess of the amount of 10.

28 Muṣāḥib (1317/1938), 160–162; Najm al-Dawla, *Jadāwil-i Lugārītm-i Aʿdād-i Ṣiḥḥāḥ az 1 tā 1000* (1292 AH/1875), 9.

confesses that although he has tried to obtain a copy of *The Spring-heads of Arithmetic* (*'Uyūn al-Ḥisāb*) by Yazdī in order to verify the specific passage, he was not able to do so. Therefore, he believes that the passage attributed to Yazdī is an addition by someone else, since there are no precedents [in the works of Islamic mathematicians] for this type of calculation.[29]

About fifty years after the publication of the first edition of Muṣāḥib's book, and after the diffusion of the idea of the discovery of the logarithm by Iṣfahānī in historical books,[30] another Iranian historian, Abu al-Qāsim Qurbānī, attacked this idea and published a note to show that this attribution is nothing more than a legend. Qurbānī pointed out that he had carefully checked the treatise of Yazdī, but had not found the phrase cited by Najm al-Dawla. Consequently, he suggested that this citation was probably an addition by one of Yazdī's readers who lived several years after the discovery of the logarithm in Europe and who had access to European logarithmic tables.[31]

The latest attempt to clarify this problem was made by an Iranian mathematics teacher, 'Abd al-Ḥusiyn Muṣ-ḥafī. He took into consideration the beginning part of Najm al-Dawla's note, according to which Yazdī had been a contemporary of "Shāh Ismā'īl" or "Shāh Sulaymān", the Safavid kings. As Muṣ-ḥafī points out, these two Safavid kings were not contemporaries. Moreover, given the fact that Yazdī had a grandson with the same first and last name, i.e., Muḥammad Bāqir Yazdī, who wrote a commentary on *Spring-heads of Arithmetic*—the very treatise of Yazdī the grandfather to which Najm al-Dawla refers—Muṣ-ḥafī suggests that the passage quoted by Najm al-Dawla might belong to the commentary on the *Spring-heads of Arithmetic*, namely *Explanation of the Difficulties of Spring-heads of Arithmetic* (*Sharḥ-i Mushkilāt-i* 'Uyūn al-Ḥisāb) by Yazdī the grandson.[32] This hypothesis becomes more plausible when Muṣ-ḥafī reminds us not only that Yazdī the grandson was a contemporary of Shāh Sulaymān,[33]

29 Muṣāḥib, Ibid., 273–274.
It is remarkable that in the second edition of his book in 1339 (SH)/1960, Muṣāḥib omitted the parts concerning the quotation from Najm al-Dawla and the discovery of logarithm by his father.

30 Cfr., for example, Humāyī (1363), 144.

31 Qurbānī (1365/1985), 139.

32 Muṣ-ḥafī, (1380/2001), 536.

33 As stated by Qurbānī, who possessed the autograph copy of *Explanation of the Difficulties of Spring-heads of Arithmetic* (*Sharḥ-i Mushkilāt-i* 'Uyūn al-Ḥisāb), the author had started to compose his treatise when "Shāh Sulaymān" was still alive, because he had written the name of "Shāh Sulaymān" in the introduction. But after completing the work in 1106/1694, he revised it by crossing the name "Sulaymān" and writing the name "Ḥusayn" above it (Qurbānī, Ibid., 440).

but also that he mentioned some European mathematical results in his commentary on the *Spring-heads of Arithmetic*.[34] As a result, as Muṣ-ḥafī presumes, the phrase quoted by Najm al-Dawla is probably due to Yazdī the grandson (in his commentary) and is based on European logarithm tables that were accessible at the time. Unfortunately, since Muṣ-ḥafī did not have access to any manuscripts of the treatises of Yazdī the grandfather or Yazdī the grandson, he could not further elaborate this conjecture. As for the attribution of invention of logarithm to Iṣfahānī, Muṣ-ḥafī assumes that since Iṣfahānī had been always eager to solve difficult mathematical problems, one can assume that he had probably discovered the function of logarithm by contemplating on this specific passage which he had eventually found in the treatise of Yazdī the grandson. Moreover, Muṣ-ḥafī suggests that Iṣfahānī had calculated a logarithm table himself which is said to be still kept by his descendants.[35]

An examination of the extant manuscripts of the *Explanation of the Difficulties of the Spring-heads of Arithmetic* by Yazdī the grandson confirms the accuracy of Muṣ-ḥafī's conjecture: the passage cited by Najm al-Dawla is found there. Fortunately, among the extant manuscripts of this treatise, there is a one kept in the Majlis library in Tehran (no. 6174/1), with marginal notes by a certain Mīrzā Naṣīr and a few glosses with the signature of Iṣfahānī at the end.[36] In addition, one of the glosses of Iṣfahānī is found in the margin of the very passage quoted by Najm al-Dawla. This gloss is as follows:

> He says: "We take the sine of twenty-five degrees."
> I say: "we take it from the table known as logarithm table, which is actually sine and tangent. The commentator did not know the truth of the table's arrangement and we should investigate the statement elsewhere."[37]

It is conceivable that this copy of the *Explanation of the Difficulties …*—preserved in the Majlis library—would be the same copy which was in the hands of Mīrzā Naṣīr, Iṣfahānī and Najm al-Dawla. However, if we assume that this

---

34 In his treatise, Yazdī, the grandson, mentions some rules in European mathematical books, according to which the amount for the number "pi" is 3,14 159 265 358 979 323 847 (MS Majlis 6174, fol. 53ʳ).

35 Muṣ-ḥafī (1380/2001), 537.

36 This signature which comports the first name of Iṣfahānī, i.e. ʿAlī Muḥammad, is the same signature which is found in the autograph copy of the anthology of Iṣfahānī (cfr. MS Majlis 81, the folios are not numbered).

37 قوله: "أخذنا جيب كه درجة." أقول: "أخذ ذلك من جدول المعروف بلكاريتم وهو ظل وجيب. ولم يطلع الشارح على حقيقة وضع الجدول ولتحقيق الكلام فيه مقام آخر." (MS Majlis. 6174/1, fol. 32ʳ)

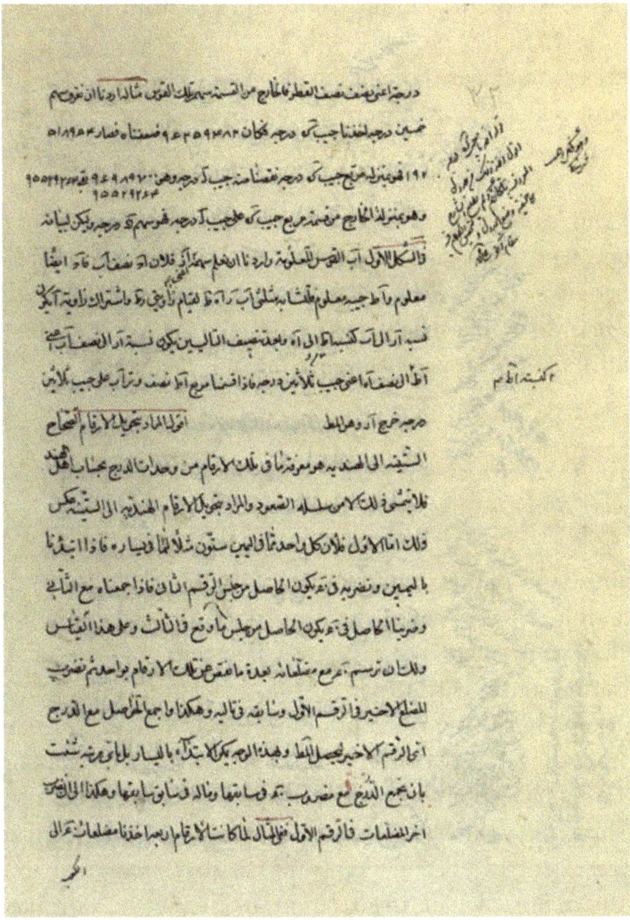

FIGURE 5.2 MS Majlis 6174/1, fol. 32[r]: The specific passage by Yazdī the grandson and the marginal note by Iṣfahānī with his signature (ʿAlī Muḥammad) at the end

"Mīrzā Naṣīr" is the same "Mīrzā Naṣīr Ṭabīb" whom Najm al-Dawla mentions, there is no marginal note of the latter ending with the phrase: "I cannot understand it" in this particular manuscript. Furthermore, the gloss of Iṣfahānī indicates that he was definitely familiar with European logarithm tables. Therefore, the evidence from this manuscript does not confirm the claim made by Najm al-Dawla concerning the invention of the logarithm by his father.

4 A General View of Iṣfahānī's Extant Works on Logarithm

In looking for the traces of the logarithm among the extant works of Iṣfahānī, two titles in the catalogues of the manuscripts held in the libraries of Iran deserve attention. The first one is entitled *Treatise on Extracting the Table of the Logarithm of the Sine from this Logarithm* (*Dar Istikhrāj-i Jadwal-i Lukārītm-i Jayb az īn Lukārītm*) and, as far as we know, it has reached us in two manuscripts. The second one, which is an anonymous treatise, is entitled *The Tables of the Logarithm in Arithmetic* (*the Science of the Numbers*) (*Jadāwil-i Lugārītm dar Ḥisāb* (*ʿIlm al-Aʿdād*)), and it has reached us in a unique manuscript. There are good reasons to attribute the latter treatise to Iṣfahānī.

5 The *Treatise on Extracting the Table of the Logarithm of the Sine from This Logarithm*

Of the two extant manuscripts of this short treatise, one of them[38] is very likely copied by Iṣfahānī's disciple, Qā'inī (see above). We attribute the copying of this manuscript to Qā'inī for two reasons. First, both the characters and the numerals are written in the same hand as Qā'inī's handwriting in his autograph extant works,[39] second, at the end of the second copy of this treatise,[40] which seems to have been copied from the first, the copyist, ʿAbd al-Ḥusayn ibn Muḥammad Mūsavī Dizfūlī, asserts that he has copied this treatise from the copy of his master, Qā'inī. Unlike the first copy, which is not dated, this second copy is dated 24 Shaʿbān 1286/ 29 November 1869. It is remarkable that there is a marginal note at the beginning of the treatise, which reads: "from the benefits of our master, Ghīyāth al-Dīn Jamshīd Thānī". For this reason, this treatise is sometimes cited among the works of Qā'inī; more precisely, it is cited as his reworking of one of Iṣfahānī's treatises.[41] But this attribution is not correct, since the last phrase of this treatise explicitly implies that the treatise is entirely from Iṣfahānī.[42]

The treatise is in three principal parts: 1) on the calculation of the sine of the degrees, 2) on the calculation of the logarithm of the sine of one degree as it

38 MS Malik 601/5.

39 Such as *A Survey of Sine and Tangent*; cfr. MS Central Library of University of Tehran 462.

40 MS Majlis 2736, fol. 166.

41 Cfr. Ḥā'irī (1350/1971), 614; ʿArshī (1394/2015), 133.

42 فالرسالة من جملة افادات سيدنا المعظم واستادنا الأنخم الأكرم، مولانا علي محمد دامت أفضاله.

has been presented "in the book" (i.e., the author's source),[43] 3) presentation of Iṣfahānī's rule in order to calculate the logarithm of the sine of one degree.

1) In the first part, Iṣfahānī explains how to obtain the sine of the degrees. In order to illustrate the method, he finds the sine of one degree, which results in 0.0174524061.[44]

2) In the second part, he talks about "the method of the author of the book",[45] according to which one needs to use the logarithms of 174 and 175 from the available table, i.e., 2.24054925 and 2.24303805 respectively,[46] in order to obtain the difference of the fractional parts; the result is 24303805 – 24054925 = 248880. Now we have to multiply 248880 by the rest of the fractional part, i.e. 524061; the result is 130427.

   To obtain the fractional part, all we need is to add 130427 to the fractional part of the logarithm of 174; the result is 24185352.

   Since the decimal part is 8,[47] the final result is 8.24185352.

   Iṣfahānī points out that this result, which is obtained by the rule explained by the author, is not accurate enough, since the number registered in the table is eighteen units more than this number.[48]

   He continues to say that he has invented a rule according to which one could obtain the result with the same accuracy as that of the number registered in the extant table.

3) In the third part, Iṣfahānī explains his own method. Instead of using the logarithm of 174, he suggests to find the logarithm of 174.5 by the method which he will later propose. He writes the results of his calculations for finding the logarithms of the numbers of the interval {174.0, 174.1, …, 174.6} in a table. Then he registers the differences between the results of every two consecutive rows in the table, as follows:

43 Iṣfahānī does not give the book's name or its author. It is probable that one of the European logarithm tables circulated among the students and professors of Dār al-Funūn, and Iṣfahānī refers to these tables.

44 Except for a few cases that the integer part of number is separated by a "comma", usually there is no decimal mark in the treatise. But in order to keep the calculations justifiable, we put the decimal mark wherever it is needed.

45 قاعدۀ صاحب كتاب

46 As we mentioned before, usually there is neither a decimal mark nor the integer part in writing the decimal numbers.

47 It is written in the margin that since the fractional part contains a number of nine digits, so the integer part is 8.

48 The table of which Iṣfahānī talks does not exist in the text. But the final result he obtains is 2418553 and he says that this is the number found in the table.

number	logarithm	difference
174.0	2.2405492500	0.0002495181
174.1	2.2407987682	0.0002493763
174.2	2.2410481444	0.0002492345
174.3	2.2412973789	0.0002490927
174.4	2.2415464716	0.0002489509
174.5	2.2417954225	0.0002488091
174.6	2.2420442316	

Iṣfahānī pursues the operation used in the previous part to obtain a more accurate approximation for the value of the sine of the logarithm of one degree: he multiplies the difference between the logarithms of 174.5 and 174.6 by the rest of the fractional part, to get

$$0.0002488091 \times 0.2406 = 0.0000598634454.$$

Adding this result to the fractional part of the logarithm 174.5, he gets

$$0.0000598634454 + 0.2417954225 = 0.2418552859$$[49]

which is more accurate than the number 0.2418553 registered in the table.

The way Iṣfahānī obtains the numbers in each row of the logarithm column is not very clear, so we can only repeat the general lines of his reasoning. Iṣfahānī considers another table, which contains the logarithms of the three numbers 174, 175 and 176. Then he calculates the two differences between these three logarithms and, at last, he calculates the difference of these two differences. He continues the calculation by multiplying the difference in the fractional part and adding the result to the logarithm of the previous number, so that he obtains the resulting number in each row.

From Iṣfahānī's explanation, we can conclude that he had at his disposal a European book on logarithms, containing tables of logarithms and a method for calculating the logarithm of a given number. The problem which faced Iṣfahānī was that when he calculated a logarithm by the method proposed by the

49 We have replaced the decimal part of logarithm of 174.5 by zero, in order to keep Iṣfahānī's calculations uniformly.

author of the book, the result he obtained was different from what already existed in the table. Therefore, he suggested a method of interpolation to calculate the logarithm with the same accuracy as that of the tables.

His method is the following: instead of taking the logarithms of 174 and 175, he divides the interval {174, 175} into 10 parts to obtain the series {174, 174.1, 174.2, …, 175}. Then he calculates the logarithms of the numbers in this series till 174.6. His method of calculation is not altogether clear, but it seems that he uses the method employed in his European source, which is probably why he does not explain it in much detail.

6 *The Tables of Logarithm in Arithmetic* (*the Science of the Numbers*)

Unlike the previous brief treatise on the logarithm, this treatise is about 280 pages in length, most of which are different types of long logarithm tables. There are also a few pages which include notes and calculations by the author, as well as some explanations on how the values in some of the tables are calculated.

As we mentioned before, this treatise, which survives in a unique manuscript,[50] is anonymous; nevertheless, we can demonstrate its attribution to Iṣfahānī with the help of three pieces of evidence:

In the first place, the handwriting of the notes scattered in the whole manuscript is the same as Iṣfahānī's handwriting in his autograph extant treatises such as *The Division of Sphere by Means of Plane Surfaces*.[51] Secondly, on the margin of page 137, the author employs a numerical example to illustrate how to calculate the sine of an arc (in a sphere) given its altitude. This numerical example and the related explanation are exactly repeated in *The Division of Sphere by Means of Plane Surfaces*.[52] The third piece of evidence comes from Iṣfahānī's son, Najm al-Dawla. In the introduction of his treatise *The Tables of the Logarithm of the Integers from 1 to 1000*, he mentions a long treatise entitled *The Science of Numbers* among the works of his father.[53] However, two titles are given on the title page of this anonymous treatise: the title is first given, in a different hand than the rest of the text as "*The Book of the Logarithm Table in Arithmetic*" (*Kitāb-i Jadwal-i Lugārītm dar Ḥisāb*) and then we see, written

50 MS Majlis 1531.
51 MS Majlis 2138.
52 Iṣfahānī, *Taqsīm-i Kura bi Suṭūḥ-i Mustawīya*, MS Majlis 2138, 4.
53 Najm al-Dawla, *Jadāwil-i Lugārītm-i Aʿdād-i Ṣiḥḥāḥ az 1 tā 1000* (1292 AH /1875), 9.

in quotation marks, "*The Science of Numbers*" (*ʿIlm al-Aʿdād*) below the former. Therefore, based on all these pieces of evidence, it is very likely that the author of this anonymous treatise is Iṣfahānī and this is probably the very treatise mentioned by Najm al-Dawla among the works of Iṣfahānī as *The Science of Numbers*. But what are all these long tables about? As an illustration of one of the simplest types of these tables, let the given numbers a and b be the logarithms of x and y respectively. Iṣfahānī explains how to apply the tables in order to find the log $(x-y)$ or log $(x+y)$ in terms of a and b. This treatise deserves an independent study.

7 Conclusion

Despite the claims made by certain authors, there is no evidence in the extant treatises and notes of Iṣfahānī to support crediting him with the discovery of the logarithm independent of European achievements. Nevertheless, as we do not have access to all of his scientific works, especially those written in Isfahan, we cannot draw a definite conclusion. What we can presume to have likely been an innovation of Iṣfahānī in the study of logarithms, is his more refined method of interpolation given in the first treatise. If we consider Iṣfahānī to be the author of the second treatise on logarithm, we can also claim that he had developed the application of the logarithm through various tables, which he evidently designed himself. Still, this seemingly small achievement shows how a traditional Iranian mathematician of the nineteenth century could elaborate on the method he has found in a European source to produce more accurate results.

8 Acknowledgement

I would like to express my great appreciation to Dr. Hossein Masoumi Hamedani not only for his support, advice and patient guidance throughout the preparation of this article, but also for the encouragement which motivated me to undertake this research. I would also like to thank Ms. Nacéra Bensaou to whom I owe the origin of my interest in ʿAlī Muḥammad Iṣfahānī and his works. I am particularly grateful to Osama Eshera, from McGill University, for his revision and comments on an earlier version of this paper. I wish to also thank Ms. Fatemeh Keighobadi, from Institute of History of Science in University of Tehran, for providing me with a copy of Najm al-Dawla's treatise on *The Tables of the Logarithm* ... I would like to thank Dr. Mohammad Bagheri for introducing me Muṣ-ḥafī's remarks in the magazine *Danesh va Mardom*. I also express my gratitude to the staff of the Majlis library (Tehran, Iran), the Central Library of the University of Tehran and Malik National library (Tehran, Iran).

Bibliography

Manuscripts

Anonymous, *Su'ālāt-i Mīrzā Jafar Khān Mushīr al-Dawlat az Ustād Mullā 'Alī Muḥammad Iṣfahānī bā Jawāb-i Ānhā*
Tehran, Majlis 1453/42.
Tehran, Majlis 81.

Anonymous, *The Tables of Logarithm in arithmetic* (*The Science of the Numbers*) (*Jadāwil-i Logārītm dar Ḥisāb* (*'Ilm al-A'dād*))
Tehran, Majlis 1531.

Ḥusaynī Qā'inī Bīrjandī Iṣfahānī, Muḥammad 'Alī (1224/1809–1311/1893), *Taḥqīq-i Jayb wa Ẓil*
Tehran, Central Library of University of Tehran 462.

Iṣfahānī, 'Alī Muḥammad (1215/1800–1293/1876), *Dar Istikhrāj-i Jadwal-i Lukārītm-i Jayb az īn Lukārītm*
Tehran, Malik 601/5.
Tehran, Majlis 2736/6.

Iṣfahānī, 'Alī Muḥammad, *Jung-i Najm al-Dawla*
Tehran, Majlis 81.

Iṣfahānī, 'Alī Muḥammad, *Taqsīm-i Kura bi Suṭūḥ-i Mustawīya*
Tehran, Majlis 2138.

I'tiḍād al-Salṭanah (Qajar minister 1276/1860–1298/1881), 'Alī Qulī Mīrzā, *Jung-i 'I'tiḍād al-Salṭanah*
Tehran, Majlis 1453.

I'tiḍād al-Salṭanah, *Tarjumi wa Sharḥ-i Bakhshī az* Āthār al-Bāqīya *Bīrūnī*
Tehran, Malik 1471.

Yazdī, Muḥammad Baqir ibn Muḥammad Ḥusiyn (eleventh/seventeenth century), *Sharḥ-i Mushkilāt-i* ʿUyūn al-Ḥisāb
Tehran, Majlis 6174/1.

Books and Articles

ʿArshī, Muḥammad-Riḍā, "Risāli-yi Nahāyat al-ʾĪḍāḥ-i Mīrzā Muḥammad ʿAlī Qāinī", *Scientific Heritage of Islam & Iran*, 2 (1394/2015), 129–162 (in Persian).

Bāmdād, Mahdī, *Sharḥ-i Ḥāl-i Rijāl-i Iran dar Qarn-i 12 wa 13 wa 14 Hijrī*, vol. 2 (Tehran: Zawwar Publication, 1378/1999) (in Persian).

Furūghī, Abu al-Ḥasan, *Awrāq-i Mushawwash yā Maqālāt Mukhtalifa* (Terhan, 1330/1951) (in Persian).

Gobineau, Joseph Arthur (1816–1882), *Trois ans en Asie (de 1855 à 1858)* (Paris: Ernest Leroux, 1905).

Ḥā'irī, ʿAbd al-Ḥusiyn, "Rīāḍīdānān-i Qarn-i 13: Mīrzā Muḥammad ʿAlī Ḥusiynī Iṣfahānī wa Rīāḍīdānān-i Hamzamān-i ū", *Waḥīd*, 91 (Tīr 1350/1971), 611–616; Reprinted in *Ḥadīth-i ʿIshq: Nukti-hā, Guftugūhā wa Maqālāt-i Ustād ʿAbd al-Ḥusiyn Ḥā'irī*, by Sahl-ʿAlī Madadī (Tehran: Library, Museum and Document Center of Iran Parliament, 1380/2001), 177–220 (in Persian).

Humāyī, Jalāl al-Dīn, *Tārīkh-i ʿUlūm-i Islāmī* (Tehran: Nashr-i Huma, 1363/1984) (in Persian).

Masoumi Hamedani, Hossein, "History of Science in Iran in the Last Four Centuries", *The Different Aspects of Islamic Culture*, vol. IV, part 2, *Science and Technology in Islam: Technology and Applied Sciences*, ed. A.Y. al-Hassan, M. Ahmed, A.Z. Iskandar (Paris: UNESCO Publishing, 2001), 615–643.

Morin, Jean-Baptist, *Trigonometriae Canonicae Libri Tres* (Paris, 1633).

Muʿallim Ḥabīb Ābādī, Muḥammad ʿAlī, *Makārim al-Āthār*, vol. 2, ed. Siyyid Muḥammad ʿAlī Ruḍātī (Isfahan: Nashri Nafāis-i Makhṭūtāt, 1364/1985) (in Persian).

Muṣāḥib, Ghulāmḥusiyn (1328/1910–1399/1979), *Jabr wa Muqābili-yi Khayyām bi Inḍimām-i Tārīkh-i ʿUlūm-i Rīāḍī az Sih Hizār Sāl Qabl az Mīlād tā Zamān-i Khayyām* (Tehran, 1317/1938) (in Persian).

Muṣ-ḥafī, ʿAbd al-Ḥusiyn, "Sar Āghāz-i Āshnāyī-i Rīāḍī-Dānān-i Īrānī bā Lugārītm", *Dānish wa Mardum*, 8 & 9, 1380 (2001) (in Persian).

Najm al-Dawla, ʿAbd al-Ghaffār (1255/1839 or 1259/1843–1326/1908), *Jadāwil-i Lugārītm-i Aʿdād-i Ṣiḥḥāḥ az 1 tā 1000* (Tehran, 1292 AH/1875) (in Persian).

Pākdāman, Nāṣir, "Mīrzā ʿAbd al-Ghaffār Najm al-Dawla wa Tashkhīṣ-i Nufūs-i Dār al-Khilāfi", *Farhang-i Iran Zamin*, 20 (1353/1974), 324–395 (in Persian).

Qurbānī, Abu al-Qāsim, "Afsāni-yi Kashf-i Lugārītm dar Iran", *Āshnāyī bā Rīāḍīāt*, vol. 8, (1365/1985), 137–139 (in Persian).

Qurbānī, Abu al-Qāsim, *Zindigī-nāmi-yi Rīāḍī-Dānān-i Duri-yi Islāmī* (Tehran: Markaz-i Nashr-i Dānishgāhī, 2nd edition, 1375/1996) (in Persian).

Rashed, Roshdi, "Mathématiques traditionnelles dans les pays islamiques au XIXe siècle: l'exemple de l'Iran", in *Transfer of Modern Science and Technology to the Muslim World*, ed. E. Ihsanoglu (Istanbul, 1992), 393–404.

Roegel, Denis, *A reconstruction of the tables of Briggs and Gellibrand's Trigonometria Britannica (1633)*, [Research Report] 2010 ⟨inria-00543943⟩.

Ṣanīʿ al-Dawla, Muḥammad Ḥasan, *al-Maʾāthir wa al-Āthār*, vol. 1, ed. Iraj Afshar, *Chihil Sāl Tārīkh-i Iran* (Tehran: Asāṭīr, 1363/1984) (in Persian).

Ṣanīʿ al-Dawla, Muḥammad Ḥasan, *Mirʾāt al-Buldān*, vol. 2, ed. ʿAbd al-Ḥusiyn Nawāyī and Mīr Hāshim Muḥaddith (Tehran: University of Tehran, 1367/1988) (in Persian).

CHAPTER 6

Formalism and Language in the Beginnings of Arabic Algebra

Marouane ben Miled

Al-Khawārizmī's *Algebra* (early ninth century) deals with equations. We will see how equations are abstract sentences that acquire meanings when they are geometrically or arithmetically interpreted. That recalls the relation between grammar and lexicography: grammar gives formalism while lexicon gives the meanings. We will use some logical tools to understand how al-Khawārizmī's *Algebra* is indebted to the works of Arabic's grammarians and lexicographers; it will allow to comprehend how al-Khawārizmī's text marks the birth of a new mathematical discipline. Then, after a summary of the beginning of grammar, we will focus on the use and the position of analogy in *Algebra*, knowing that, in grammar, it has a central position to establish rules, to justify them and to build new words.

1 Some of al-Khawārizmī's *Algebra*

In his *Algebra*, al-Khawārizmī begins by setting three kinds of "numbers" (*aʿdād*): *al-māl* (the Arabic word for money or goods); the radix (*al-jidhr*) and the simple number (*al-ʿadad al-mufrad*). Simple numbers are a class of numbers. They are not opposed to some composed numbers, but to the *māl* and the radix. The numbers classically considered by the arithmeticians, outside the framework of algebra, could be formalised by one or the other of the three primitive terms called "numbers". He sets that the *māl* is the **square** of the *radix*; the *radix* being "the **thing** [the product of which] [...] is the *māl*". Let's yet say, to help the reader, that according to our modern representation *al-māl* corresponds to x^2; the radix corresponds to x and the simple numbers correspond to our class of positive integers (sometimes rationals). We could also translate the *māl* by x, and the *jidhr* by $\sqrt{x}$.

Al-Khawārizmī uses these kinds of "numbers" as three primitive terms, to build sentences, through a reasoning exploring all the possible combinations with an addition and the relation of equality. He obtains six sentences on the following patterns: *māl*s are **equal** to *radices* ($ax^2 = bx$); *māl*s are **equal** to the

 | DOI:10.1163/9789004513402_008

[*simple*] *number* ($ax^2 = c$); *radices* are **equal** to the [*simple*] *number* ($bx = c$); *māls* **plus** *radices* are **equal** to the [*simple*] *number* ($ax^2 + bx = c$); *māls* **plus** the [*simple*] *number* are **equal** to *radices* ($ax^2 + c = bx$); *radices* **plus** the [*simple*] *number* are **equal** to *māls* ($bx + c = ax^2$).[1]

So, these canonical equations are built by a syntactical procedure. They are built out of any semantic: equations are neither true nor false, more than that, they contradict each other, *they are not propositions*, and therefore, *they cannot be proved*. It is important to underline that it is something completely new in mathematics, to deal with sentences that are not propositions; this is the most important point, from which all algebra will develop.

For each of these six equations, al-Khawārizmī gives an algorithm which permits to find the positive solution(s) when it (they) exist(s). For example, the equation "*māl* plus radices are equal to numbers" ($x^2 + bx = c$), has as algorithmic solution $x = \sqrt{((\frac{b}{2})^2 + c)} - \frac{b}{2}$.[2]

Each algorithm is *validated* arithmetically or geometrically by a figure and algebraic-geometrical techniques.[3] It is here, and only here, that meanings are given to the equations and to their resolutions' algorithms by the *interpretations* of the equality, the sum and the three terms *māl*, radix and simple numbers.[4] In other words, we deal with a semantisation obtained by the interpretation of the terms "*māl*", "radix" and "simple numbers", the addition and the equality, in one or the other of the arithmetical or geometrical framework. That is how the algorithms are validated. Nowadays, we speak about proofs of correctness of the algorithms. We have now to understand how these interpretations are done.

2 The Interpretation

2.1 *The Interpretation of Terms*

The *māl* is arithmetically interpreted as a squared number, and geometrically as a square. The radix is arithmetically interpreted as a number (which eventually could be squared), and geometrically as a straight line segment (which eventually could be the side of a square). Simple numbers are arithmetically interpreted as natural numbers and geometrically as (measured) magnitudes.

---

1 The classification is made according to the number of terms and not the degree. The degree will be introduced later by al-Karajī (tenth-eleventh centuries), Ben Miled (2004).

2 Cfr. Rashed (2007), 108–120.

3 Cfr. Rashed (2007), 100–120.

4 Cfr. Ben Miled (2015).

2.2 *The Interpretation of the Formal Sum*

In the text, we deal with the additive conjunction between terms, named by the letter *wāw*, the literal meaning of which is "and". Its arithmetical interpretation is the sum of two numbers. Its geometrical interpretation is the sum of two segments or of two surfaces (adjunct of a segment to another; construction of a surface which is equal to the sum of two given surfaces). It turns out that the terminology used by the author is precise: in the formal framework, the sum is always denoted by the letter *wāw*, it is the additive coordinating conjunction, as I said. In the interpreted framework, it is named by the verb *zāda* (to add).

2.3 *The Interpretation of the Formal Equality*

In the text, the author uses the verb *ʿadala*, which literally means "to equal". Its arithmetical interpretation is the identity between two operating procedures on numbers, or between the operating procedure and its result. Its geometrical interpretation is the identity between two geometrical procedures (constructions) which ends by the construction of the same magnitude. So, the terminology of the author is clear: in the formal framework, the equality is systematically named either by the verb *ʿadala* or by the verb *ṣāra* (to be, to become), while in the interpreted framework the word *huwa* (fem. *hiya*), which expresses the relation of identity, is used.

3 The Formalisation

Reciprocally, we have to study how arithmetical and geometrical problems are formalised.

3.1 *Formalisation of the Arithmetical Problems*

If the problem consists of finding or knowing an "unknown" (*majhūl*) number from data, this unknown number will be put as a "thing" (*shayʾ*) or a "radix" (*jidhr*), and its square, if it appears in the problem's data, or in its resolution, is put as being the "*māl*". The other numbers are "simple numbers".

3.2 *Formalisation of the Geometrical Problems*

The formalisation of the geometrical problems was done first by al-Khawārizmī and then, massively, by his successors.[5] If the problem consists in **knowing** an "unknown" linear magnitude, this magnitude is **put** as a "thing" (*shayʾ*) or the

5 Cfr. in Ben Miled (2005), Farès (2016) and Ben Miled (1999) how the geometrical propositions

radix (*jidhr*); and it square, if it appears in the problem's data, or in its resolution, is **put** as the "*māl*"; the other magnitudes are "simple numbers".

Let's here focus on the verbs "to know" and "to put" that were emphasised above. "To know" is the translation of the verb *'arafa* used here by al-Khawārizmī. Knowing a magnitude presupposes to measure it,[6] and consequently has to be considered as the algebraic object which formalises an irrational magnitude as a "simple number". The algebraical object "radix of two" which formalises an irrational magnitude, is the "the thing which *māl*" is equal to two, because the "radix is each thing multiplied by itself" and the *māl* is what results.[7] "To put" is the translation of the verb *ja'ala*[8] used here by al-Khawārizmī. It makes a substitution by which the problem, with an unknown parameter, is formalised in a problem on the formal "unknown", which is a fixed term of algebra independent of the problem. Moreover, we observe that the same arithmetical or geometrical magnitude can be formalised as one or the other of the three terms of algebra, depending on the problem. Hence, *it is the situation of the magnitude in the problem's statement which determines its syntactical function during the formalisation.*

Reciprocally, more than one interpretation is possible for a same form. For example: we have two different geometrical constructions which validate the algorithm of resolution of the equation $x^2 + bx = c$.[9] In some parts of the text the interpretations decide which of the two solutions of the equation ($ax^2 + c = bx$) we have to take.[10] In others, when the discriminant is negative, there is no possible interpretation for an equation: "the problem is impossible"[11] writes al-Khawārizmī (the problem! Not the equation).

Algebra, as elaborated by al-Khawārizmī, is an abstract language, founded on a syntactical construction. It appears as a place, where, thanks to formalisation, geometrical and arithmetical concepts, constructions and propositions meet. Al-Khawārizmī's algebra acts as a common empty language in which both arithmetic and geometry are expressed; it is a formal bridge between their semantic frameworks. *It can potentially be interpreted in any model that has a relation of equality and an operation of addition* … and it will be. We can consider equations with integers coefficients on any ring.

---

of the Tenth's Euclid's *Elements* are interpreted in the terms of the *Algebra* of al-Khawārizmī. Cfr. in Diophante (1984) how Diophantus' problems were formalised.

6 Cfr. Ben Miled (2008).

7 Cfr. Rashed (2007), 97.

8 Cfr. Rashed (2007), 147–149.

9 Cfr. Rashed (2007), 109–121.

10 Cfr. Rashed (2007), 105.

11 Cfr. Rashed (2007), 107.

4 New Mathematical Objects Are Created

The abstract unknowns are new mathematical objects that al-Khawārizmī and his followers have called "numbers". On these "numbers" were defined operations, expressed properties, introduced an equality. This operating practice on formal entities contributed to give them a meaning, as it must have been the case a few millenniums earlier (maybe in the same area) with the mathematical ideals that are the numbers and the geometrical magnitudes. That makes algebra a new domain with new mathematical objects. Their meanings are independent from the interpretations. They are generated by the formal proofs (calculations), the formal rules, the relation of equality between formal expressions, the quantification on the variables, etc.

For example, the normalisation of an equation ($ax^2 = c \Rightarrow x^2 = \frac{c}{a}$), contributes to give sense to the formal equality: "All the *māls*, the positive and the negative, are returned to one *māl* ... We proceed the same for the numbers which are equal".[12]

Let us first note that syntactically, an equality between two identical terms is trivial, and has no interest and is not treated by al-Khawārizmī. In an equality between different terms, as in the expression "*māls* are equal to the [simple] number", the formal transformation of this equation into another, for example by normalizing it, makes it possible to constrain formal equality, and thus helps to give it meaning, by the very fact that this constraint is verified only by equality within the geometric or arithmetic model.

In the same way, in the following example, we see how free variables are universally quantified, and how the universal quantification on the *māl* contributes to give sense to the variables, to the sum on them and to the equality between them:

> Two *māls* and ten radices are equal to forty-eight drachmas. That means that *any*[13] two *māls* (*'ayyi mālayni*), if they are summed and added to ten radices of one of them, it results forty-eight drachmas.

Of course, universal quantification is not defined as such, it is by the determination of all instantiations of the variable through the determinant "any", it thus links formal operations (here the formal addition) with formal equality, contributing once again to set their limits and to define them.

12 See [Rashed (2007), 99].

13 I'm emphasizing.

So, we have seen how the two reciprocal steps of formalisation and of interpretation take a central place in the new algebra developed by al-Khawārizmī. That leads to question how we can make any formal proof on formal rules. Certainly not as Euclid (300 BCE) did in the *Elements*, because an axiomatisation of the algebra is still lacking. We saw that the legitimisation of the formal rules are given by their geometrical, sometimes arithmetical, interpretations which give them a meaning and so, justify them. Later, al-Karajī will give a first solution to this problem in *'Ilal ḥisāb al-jabr wa al-muqābala* (*The Causes of Calculation in Algebra*).[14]

In conclusion: Al-Khawārizmī treats in a differentiated way the syntactic and semantic aspects, with a special lexicon for each one. He constructs forms, which are not proposals, on two abstract terms (the radix and the *māl*), independently of the problems. The meanings are given through the interpretation of the terms, the formal operations and the formal equality, geometrically or arithmetically. In a movement towards a still higher level of abstraction, he presents some formal rules on an indeterminate variable subject to universal quantification.

We see that an algebra—in the sense of an abstract language, allowing to build forms, and initiating a calculation on it, lending itself to different interpretations—does not require the recourse to a symbolic writing, it can be established in a natural language; it is enough to compose from a given number of words, empty of semantic content: here it is done by the two words *radix* and *māl*, but it will very quickly done by the word *shay'* (thing), through which the various tiered constructions of its powers are defined as well as their relation of proportionality [Rashed (1975), Ben Miled (2004)]. The term "thing" being by definition meaningless, it is subject to all interpretations, and constitutes the indeterminate variable that will allow the development of algebra and its emancipation from geometry. That reminds the relation between grammar and lexicography: grammar gives a formalism while lexicon gives the meanings.

Before exploring further the relationship with grammar, it is interesting to recall its genesis.

---

14 See [Nafti (2018)] and [Ben Miled and Nafti (2019)].

5 Beginnings of Arabic Grammar

5.1 *Out of the Arabian Island*

With the early Muslim conquests, the Arabic language left the Arabian Peninsula to become the official language of the Umayyad Caliphate (661–750) "from the Atlantic Ocean to the confines of India" as the expression goes. More than a religious language, it became the language of the power and of laws, as well as of culture, sciences and philosophy. Then the Persians and the Greeks and other peoples of the conquered areas of both the Sassanian and the Byzantine empires, adopted Arabic.

That explains the known movement of translations from Greek to Arabic; directly, such as for Euclids' *Elements*, by Ḥajjāj b. Maṭar,[15] under the Caliphates of Harūn al-Rashīd (786–809) and al-Ma'mūn (813–833), or via the Syriac, as for Aristotle's *Categories*.[16] Let us precise that the translations were not neutral, as it can be seen from the choices of the texts to be translated and from the introduction of new concepts and methods in relation to the developments of philosophy and science.[17]

5.2 *Necessity of Founding a Grammar, and the Reasoning by Analogy*

Non-native Arabic speakers had to conceive grammar for the Arabic. Two schools, among others, emerged in the eighth century: one in Kūfa and another in Baṣra with al-Khalīl (eighth century) and al-Sībawayh (end of the eighth century).[18]

Let us quote to begin that the reasoning by analogy is central in Grammar. Gérard Troupeau writes:[19]

> La science dẹs fondements de la grammaire: c' est l' étude des arguments (*adilla*) généraux de la grammaire: la transmission auditive (*samāʿ*), le consensus (*ijmāʿ*) et le raisonnement par analogie (*qiyās*).

And, about the analogical method of Baṣra, Chartouni and Grand'Henry write:

> [La grammaire] a pour fondement que la langue étant créée par Dieu comme le reste du monde, elle est un reflet de logique, donc de 'l' Intelli-

15 Cfr. De Young (2002).
16 Cfr. Hugonnard (2004).
17 Cfr. Rashed (2017), XXXV–LIV.
18 Cfr. Troupeau (EI).
19 In Troupeau (1995).

> gence et de la Perfection divines'. Établir des règles de grammaire consiste donc à retrouver dans la langue des expressions de cette Perfection divine. En morphologie, tout le système se développe par le principe d'analogie (*qiyās*) et similitude (*tashākul*).[20]

We can also refer to Ibn al-Jinnī (ninth century) who writes: "The power of the analogy is that for the grammarians, what is analogous to a statement of the Arabs, is Arabic".[21]

As in grammar, analogy is also a fundamental tool in juridical science. Elamrani-Jamal writes:

> Cet instrument essentiel dans l'histoire des sciences linguistiques et juridiques dans les trois premiers siècles [seventh to ninth centuries], suscita les controverses que l'on connaît entre partisans et adversaires de son usage en grammaire, l'École de Baṣra contre l'École de Kūfa, déjà dans le *fiqh* entre malékites et chafiites d'une part, et Hanbalites de l'autre.[22]

5.3 *Grammar and Mathematics*

Roshdi Rashed showed how al-Khalīl used combinatorial mathematical results to produce all the possible entries of a dictionary, submitting them to a phonological study;[23] and I have given a new enlightenment in the first part of this paper that links grammar and the first algebraic text. Let's now discuss the existence of a rule of *qiyās* (analogy) in al-Khawārizmī's *Algebra*.

6 Is There Any *qiyās* in al-Khawārizmī's *Algebra*?

The word *qiyās* appears very frequently in the text, always in the meaning of a rule or a method to apply to a situation. But, all the rules are not *qiyās*. For example, al-Khawārizmī uses the word *qiyās* for an analogy which treats all the similar cases as a first given one, which is a general-example; so, here, the *qiyās* is a *generalisation*.

> [To] double the radix of any *māl*, known or surd, you have to multiply two by two and by the *māl*, the radix of the result is twice the radix of the *māl*.

20 In Chartouni (2000), 14.
21 In Elamrani (1983), 44]. Translation is mine.
22 In Elamrani (1983), 44.
23 Cfr. Rashed (2014), 149– …; Rashed (2007).

> If you want to multiply [the radix of any *māl*] by three, multiply three by three and by the *māl*, the radix of the result is three times the radix of the *māl*. You do the same for the other multiples, on this example you make the analogy (*qiyās*).[24]

In some other parts of the text,[25] al-Khawārizmī uses the word *qiyās* for a method which consists of translating the terms of a given problem into the terms of algebra: it is a *formalisation*.

> First problem of six: When you say, ten is divided into two parts, you multiply one by the other, and you multiply one of both by itself, in such way the multiplied by itself is as one part by the other four times. Its *qiyās* is to let one of the two part a thing, and the other ten minus a thing, and to multiply …[26]

Here, the *qiyās* is the systematic translation of the terms of the problem into the abstract terms of algebra. If there is any analogy, it is made with a phantom model which is not given (a general rule is given instead).

A quasi-analogy which is not *qiyās* is given for calculations on the formal expressions i.e. which contain the term "thing":

> I'll teach you how to multiply one to the others the *things*, which are the radices, if they are alone; if they are with a number; if they are decreased by a number; or if they are subtracted from a number; and how to add them one to the others; and how to subtract them one from the other.[27]

Al-Khawārizmī gives what he calls, "proofs by indication" (*dalīl*): he writes: "I showed you (*bayyantu*) that to **indicate** to you (*li tastadilla*) how to multiply".[28] The "proofs by indication" are not by analogy (*qiyās*). For example: the rules of signs and the development of forms of the type $(ax + b)(b' - a'x)$ (where x designs "the thing") are given on the model of calculations of the same type, where 10 is substituted to x. So we deal here with the justification of a formal rule, by the truthfulness of its interpretation in the arithmetical

24 In Rashed (2007), 131, l. 17.
25 Cfr. Rashed (2007), 145–159.
26 In Rashed (2007), 145, l. 10.
27 In Rashed (2007), 122.
28 See Rashed (2007), 124–125.

model. Al-Khawārizmī says that calculations on numbers are an "indication" for the calculations on forms with "the thing". It isn't an analogy because the analogy should be done between similar problems expressed in the same language; here, from a model of calculation on numbers (natural numbers), we deal with a calculation on abstract entities.

Reciprocally, applying a general rule is not making an analogy, because al-Khawārizmī proceeded by a combinatory method that placed him immediately at a level of higher abstraction; the procedure of abstraction or formalisation (translation in the terms of algebra) of arithmetical and geometrical problems ensues.

We see that the proofs by *qiyās*, in al-Khawārizmī's text, are not all of the same kind. It would be interesting to compare the use of *qiyās* in the works of the Muʿtazilites and the theologians of the same period, such as Abū Yaʿqūb al-Shaḥḥām, as it would be interesting to compare the significance of the word *shay'* in both the mathematical and the theological works; it would also be enlightening to follow the use of *qiyās* in the texts of Abū Kāmil (ninth-tenth centuries),[29] Sinān ibn al-Fatḥ (tenth century) etc.[30] as a step forward in the research into the techniques of algebraic proofs, until the reform of al-Karajī.

Bibliography

Ben Miled, Marouane, "Formes et sens dans l'*Algèbre* d'al-Khawārizmī", in *Histoire et Philosophie des Mathématiques, et disciplines associées en Méditerranée*, presentation given in november 2015 (Marseilles: Presses universitaires de Provence, forthcoming).

Ben Miled, Marouane, "Formalism and language in the beginnings of Arabic algebra", presentation given at the Symposium 110: "Scientists and the powerful from the Middle Age to the Classical period" (Prague: 7th ESHS Conference, Science and power, science as power, 22 of Sept. 2016).

Ben Miled, Marouane Ben Miled. "Mesurer le continu, dans la tradition arabe des Livre V et X des *Éléments*", *Arabic sciences and philosophy*, 18:1 (Cambridge: Cambridge University Press, 2008), 1–18.

Ben Miled, Marouane, *Opérer sur le Continu, traditions arabes du Livre X des* Éléments *d'Euclide* (Carthage: The Tunisian Academy of Sciences, Letters and Arts *Beït al-Hikma*, 2005).

29 Cfr. Rashed (2012).

30 Cfr. Rashed (1975), 33–60 or Rashed (1984), 43–70 and Nafti (2018).

Ben Miled, Marouane, "Les quantités irrationnelles dans l'œuvre d'al-Karajī", in Régis Morelon and Ahmad Hasnawi (eds), *De Zénon d'Élée à Poincaré* (Leuven-Paris: Peeters, 2004), 27–54.

Ben Miled, Marouane, "Les commentaires d'al-Māhānī et d'un anonyme du Livre X des *Éléments* d'Euclide", *Arabic sciences and philosophy*, 9:1 (1999), 89–156.

Ben Miled, Marouane and Nafti, Foued, "Causes du calcul algébrique d'al-Karajī, présentation et analyse", in *Proceedings of the 10th international colloquium of the SIHSPAI, Science, philosophy and kalam in Islamic civilisation: the old and the new* (Napoli: 9–11 sep. 2019), forthcoming.

De Young, Gregg, "The Arabic version of Euclid's *Elements* by al-Ḥajjāj ibn Yūsuf ibn Maṭar", *Zeitschrift für Geschichte der arabisch-islamischen Wissenschaften*, 15 (2002), 125–164.

Diophante, *Les Arithmétiques*, Livre IV (tome III), Livres V–VI–VII (tome IV), ed. Roshdi Rashed (Paris: Les Belles Lettres, 1984).

Elamrani-Jamal, Abdelali, Logique aristotélicienne et grammaire arabe: étude et documents (Paris: J. Vrin, 1983).

Farès, Nicolas, *Commentaire du Livre X des* Éléments *d'Euclide par Abū Ja'far al-Khāzin* (Beyrouth: Publications de l'Université libanaise, 2016).

Hugonnard-Roche, Henri. *La logique d'Aristote Du grec au syriaque* (Paris: Vrin, septembre 2004).

Al-Khwārizmī, Muhammad Ibn Mūsā (early nineth century), *Kitāb al-jabr wa-al-muqābala*, ed. Roshdi Rashed, *Le commencement de l'algèbre* (Paris: Librairie Scientifique et Technique Albert Blanchard, 2007).

Nafti, Foued, *Al-Karajī rénovateur de l'algèbre, avec l'édition critique de 'Ilal ḥisāb al-jabr*. PhD thesis, (Tunis: École nationale d'Ingénieurs de Tunis, juillet 2018).

Rashed, Roshdi (ed.), *Lexique historique de la langue scientifique arabe*. (Hildesheim Zürich New York: Georg Olms Verlag, 2017).

Rashed, Roshdi, *Classical Mathematics from Al-Khwārizmī to Descartes* (New York: Routledge/Taylor & Francis Group, 2014).

Rashed, Roshdi, *Abū Kamil: Algèbre et analyse diophantienne. Édition, traduction et commentaire* (Berlin-New York: Walter de Gruyter, 2012).

Rashed, Roshdi, *Entre Arithmétique et Algèbre* (Paris: Les Belles Lettres, 1984).

Rashed, Roshdi, "Recommencement de l'algèbre aux XIe et XIIe siècles", in J.E. Murdoch and E.D. Sylla (eds.), Cultural Context of Medieval Learning, Proceedings of the First International Colloquium on Philosophy, Science, and Theology in the Middle Ages, September 1973 (Dordrecht—Holland: Reidel Publishing Company, 1975), 33–60. Taken up in Rashed (1984), 43–70.

Rashed, Roshdi, "Algèbre et linguistique: l'analyse combinatoire dans la science arabe", in *Boston Studies in the Philosophy of Sciences* (Boston: Reidel, 1973), 383–399.

Chartouni, Rachid and Grand'henry, Jacques. *Grammaire arabe à l'usage des Arabes:*

traduction et commentaires des Éléments d'arabe, morphologie et syntaxe (Leuven: Peeters, 2000).

Troupeau, Gérard, "Naḥw", in *Encyclopédie de l'Islam*, 2nd edition (Leiden: Brill, 1954–2005).

Troupeau, Gérard, *Philologie arabe*, 4e section, sciences historiques et philologiques, Livret 11 (Paris: École pratique des Hautes études, 1995–1996, 1997), 42–43.

CHAPTER 7

Art and Mathematics, Two Different Paths to the Same Truth

Patricia Radelet-de Grave

As my first image shows, I would like to invite you into the wonderful world of Arabic pictorial art. As you immediately feel it when looking at that picture, this pictorial world is completely different from ours, principally because of the absence of human representations, at least in this picture and as well as in the pictures we shall consider. In other regions, pictorial art is on the contrary principally centred on human representations so that Arabic art gave rise to a totally different research and led to a new knowledge, that I would like to introduce to you.

One may look at Arabic pictorial art from different points of view. Emile Prisse d'Avennes (1807–1879) classified those designs from the point of view of the places they decorate, Owen Johns (1809–1874) on the contrary grouped them by forms and colours. They both tried also to classify them. I used their collections of pictures, as Andreas Speiser (1885–1970) and Hermann Weyl (1885–1955) had already done it. Speiser's principal aim was to perfect their classification using mathematical concepts. That method of classification as well Weyl's work on symmetries played an important role in the birth of group theory at the beginning of the twentieth century. The classification is based on the symmetries that organise the picture and led to a crucial notion in physics today, namely "invariance". Today, this kind of mathematics is part of common mathematics.

Andreas Speiser and Hermann Weyl with many others contributed to the elaboration of group theory and of the notion of invariance. Both mathematicians were well aware of the debt they had to Arabic art. Andreas Speiser writes in the introduction to his work on *Theory of groups of finite order*, 1923 (1956):

> 1. About the prehistory of group theory
> Long before permutations were studied, mathematical figures were constructed which were closely linked to the theory of groups, and which could only be grasped by means of concepts derived from this theory, such as regular patterns, which by displacements or by specular images, cover

 | DOI:10.1163/9789004513402_009

FIGURE 7.1 Arabic pictorial art. Jones (1856), plate XXXV

> surfaces. They constitute, with music, one of the high facts of the high mathematics of Antiquity.[1]

and he concludes a page further:

> Unfortunately, the Egyptian and Arabic ornaments have never been studied according to their geometrical content, and so one of the most beautiful chapters in the history of mathematics remains to be written.[2]

Owen Jones was an architect and designer who played a role in the elaboration of the modern theory of colours and elaborated the first tentative theories about flat patterning and ornament. On the title page of his book on the Alhambra (1842–1845) he writes:

---

1 Speiser, 1923 (1956), p. 1. I. Zur Vorgeschichte der Gruppentheorie
Lange bevor man sich mit Permutationen beschäftigte, wurden mathematische Figuren konstruiert, die auf das engste mit der Gruppentheorie zusammenhängen und nur mit gruppentheorischen Begriffen erfasst werden können, nämlich die regulären Muster, welche durch Bewegungen und Spiegelungen mit sich selbst zur Deckung gebracht erfasst werden können. Sie bilden zusammen mit der Musik einen Hauptgegenstand der höheren Mathematik im Altertum.

2 Speiser, 1923 (1956), pp. 2–3: Leider sind die ägyptischen und arabischen Ornamente bisher nie nach ihrem geometrischen Gehalt untersucht worden, und so bleibt eines der schönsten Kapitel der Geschichte der Mathematik noch zu schreiben.

FIGURE 7.2 Arabic screen
PICTURE BY THE AUTHOR, © KURT SCHMIDT

> The Alhambra Palace, that geniuses have painted gold as dreams and filled with harmonies.[3]

Using the word "harmonies", Owen Jones makes, as Speiser did in the previous quotation, a link with music.

3 La Alhambra palais que les génies ont doré comme un rêve et rempli d'harmonies, Jones, 1842–1845, Title page.

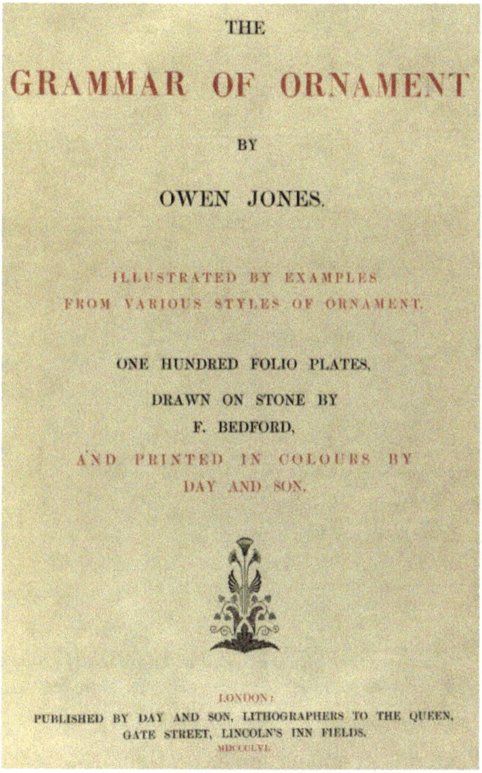
THE

GRAMMAR OF ORNAMENT

BY

OWEN JONES.

ILLUSTRATED BY EXAMPLES
FROM VARIOUS STYLES OF ORNAMENT.

ONE HUNDRED FOLIO PLATES,

DRAWN ON STONE BY

F. BEDFORD,

AND PRINTED IN COLOURS BY

DAY AND SON.

LONDON:

PUBLISHED BY DAY AND SON, LITHOGRAPHERS TO THE QUEEN,
GATE STREET, LINCOLN'S INN FIELDS.
MDCCCLVI.

FIGURE 7.3
Owen Jones, *The grammar of ornament*, 1877

Hermann Weyl writes in his book *Symmetry* (1952) on p. 109:

> The greatest masters of the geometric art of ornament were the Arabs. The walls of such buildings of Arabic origin as the Alhambra in Granada is simply overwhelming.

These quotations make clear that those who participated in the elaboration of group theory were well aware of the importance Arabic art had for understanding the theory they were shaping. And this despite the enormous distance in time between them and their masters.

From this point of view, books by historians of art as Owen Jones and Emile Prisse d'Avennes played an important role because they offered lots of examples that could be analysed from the mathematical point of view. Andreas Speiser mentions them both.

The aim of this article is to show how mathematicians used those ornaments to develop group theory. But to do this, I have first to explain what is invariance or symmetries, the central mathematical idea of those ornaments.

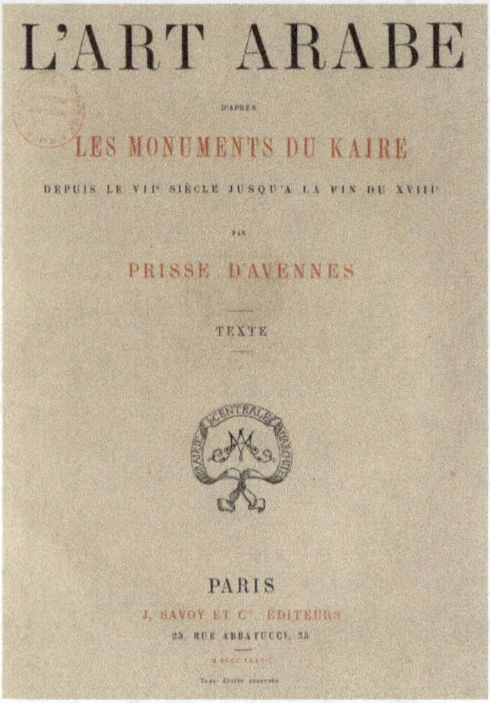
L'ART ARABE

D'APRÈS

LES MONUMENTS DU KAIRE

DEPUIS LE VIIe SIÈCLE JUSQU'À LA FIN DU XVIIIe

PAR

PRISSE D'AVENNES

TEXTE

PARIS

J. SAVOY ET Cie, ÉDITEURS

25, RUE ABBATUCCI, 25

FIGURE 7.4
Emile Prisse d'Avennes, *L'art arabe*, 1856

To speak of invariance, you need an object and a transformation. Invariance says that the object is invariant under that transformation. Here the objects are the motives, elements of the ornament and they will be transformed. In this case, the transformations will be displacements, rotations or reflections in the plane of the ornament. Those motives have to be invariant, they must remain the same after a displacement, a rotation or a reflection. That is the way Arabic artists constructed their ornaments. And this is also the definition of invariance in physics today. But in physics, the concepts of transformation and of motive are much more general as I shall show in my article on the prehistory of relativity. In fact "invariance" is a very general notion, in the case of ornaments I should speak of symmetries.

1 Mathematical Theory of Symmetries in One, Two and Three Dimensions and Art

1.1 *In One Dimension: Friezes and Borders*

Friezes are not strictly one-dimensional. They are strips with a certain width that is a restriction of two-dimensional ornaments. Their particularity is a central line called main axis that divides longitudinally the frieze. Mathematically one doesn't consider friezes particularly, but they will help me to explain some properties on more simple objects.

1.1.1 Transformations

In friezes you may observe four different transformations, displacements or shifts. Namely[4]

1) Translations along the main axis of the frieze.
2) Reflections or shifted reflections with respect to the main axis of the frieze.
3) Reflections with respect to axis perpendicular to the main axis of the frieze.
4) Rotations of 180° around any point of the main axis.

You may also combine those transformations to get more elaborated ornaments. Let us see how it works using some examples.

1) Purely translational symmetry

In one dimension the idea is very simple, you take one "element" or a motive and you repeat it many times along one line. You may also take a cell or a fundamental region that contains the motive. Those cells or fundamental regions are also characteristic of the symmetry. They can be triangles, squares, rectangles or parallelograms. Andreas Speiser illustrates this in the following way:[5]

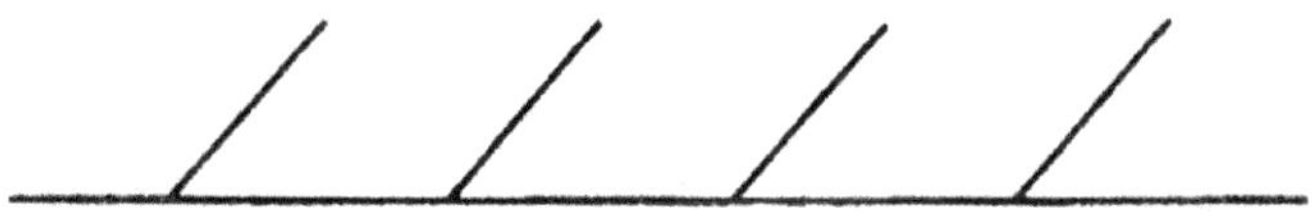

FIGURE 7.5A

FIGURE 7.5

He takes a very simple element without any symmetry, an oblique line and shifts it along a guideline, the main axis of the frieze.

4 Speiser, 1923 (1956), p. 81.
5 Figures and inspiration are taken from Speiser, 1923 (1956), pp. 81–82.

FIGURE 7.6 Prisse d'Avennes (1869–1877) gives several examples of friezes put together on page 139.

Arabesques: mosque of Ahmed-ibn-Touloun, details of ornamentation (ninth century)

2) Reflection with respect to the main axis

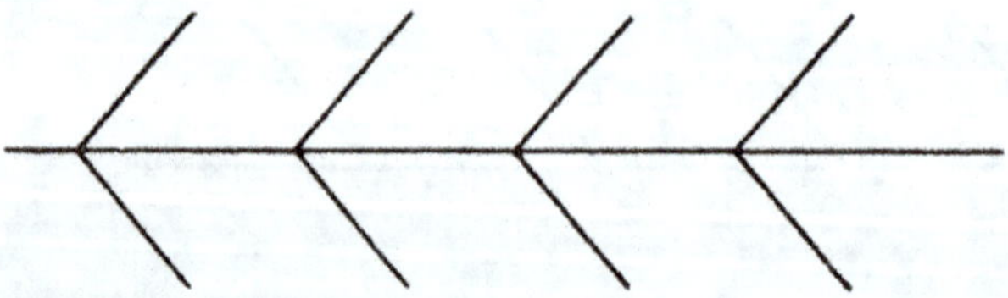

FIGURE 7.7A

FIGURE 7.7

FIGURE 7.8

3) Reflection with respect to axes perpendicular to the main axis

FIGURE 7.9A

We find several examples of that symmetry on Prisse d'Avennes' plate, here are two of them.

FIGURE 7.9

FIGURE 7.10

4) Rotation of 180° around axes perpendicular to the plane of the ornament

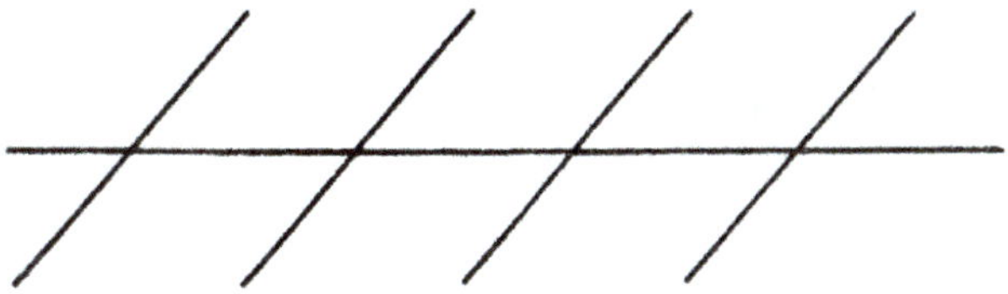

FIGURE 7.11A

FIGURE 7.11

The example taken from Prisse d'Avennes (1869–1877), p. 153 is an Arabic ornament elaborated between the fifteenth and the eighteenth century.

5) The ornament has the three precedent symmetries together.

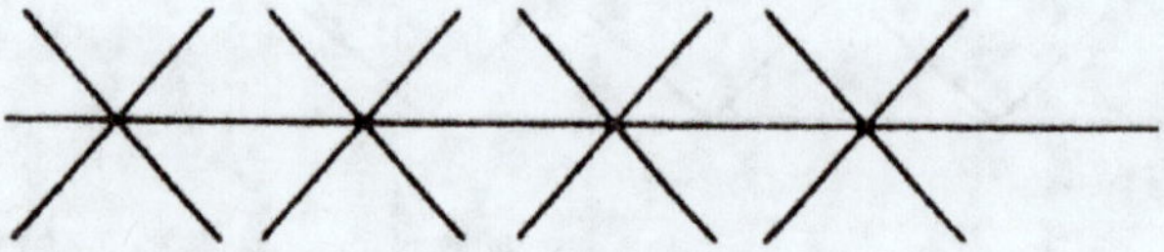

FIGURE 7.12A

FIGURE 7.12

Here is an example taken from another plate of Prisse d'Avennes' (1869–1877), p. 223 left.

6) Translation and reflection with respect to the main axis (the main axis is a glide axis)

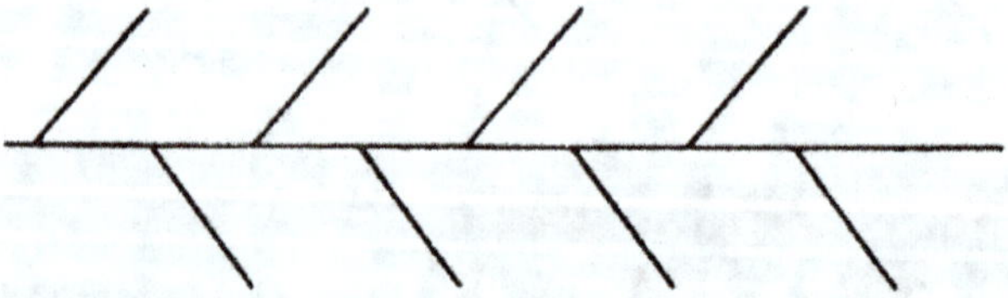

FIGURE 7.13A

FIGURE 7.13

FIGURE 7.14

Here we go back to Prisse d'Avennes' samples. We can see the difference between example 4 and this one. The element is not only turned but is also shifted.

7) The ornament has the symmetries 3, 4 and 6 together.

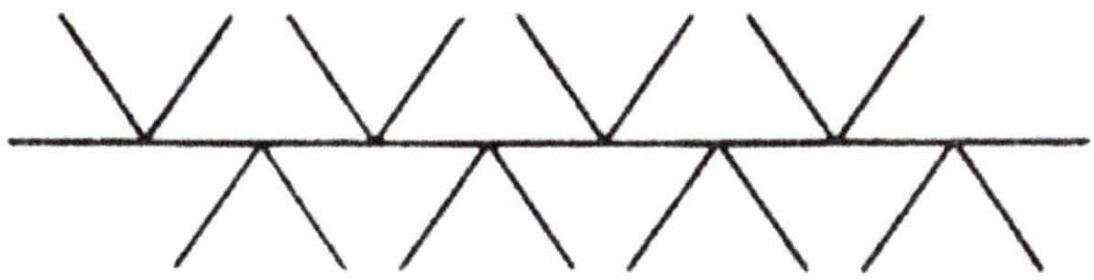

FIGURE 7.15A

FIGURE 7.15

FIGURE 7.16

On the frieze on the right, remark that the little frieze under the principal one, has the same symmetry but a completely different motive. This is noteworthy!

There are only seven different possible friezes or borders but you may change the motive if you keep its symmetry. This makes the richness.

On the next two examples given by Owen Jones, you can see that colour may reduce or even destroy the symmetry[6]

FIGURE 7.17
Arabian ornaments Owen Jones (1856) on plate XXXV, nr 28 and nr 18

FIGURE 7.18

6 Jones, figs. 18 and 28 on plate XXXV.

1.2 *Two Dimensions: Wallpaper's Group*

Historians usually attribute to George Polya (1887–1985) the first demonstration (1924) of the fact that there are only 17 different patterns, from the mathematical point of view, in two dimensions.[7] Here is the table he gives.

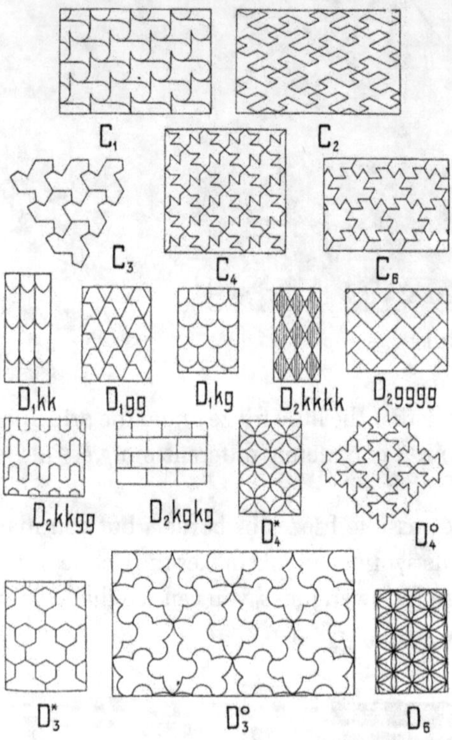

FIGURE 7.19
Polya's characterisation of the 17 patterns in two dimensions

In fact, that result had already been demonstrated by Fricke and published in 1897 together with Felix Klein.[8] Some years before Polya, Schönflies[9] and Fedorov[10] had demonstrated independently that there are 230 different patterns in three dimensions. But this concerns crystallography and I shall not enter into that problem.

Let us concentrate on two dimensions.

There is a notion I didn't mention yet: the lattice. The lattice has to do with the motive, it is composed of repeated elementary cells that contain the

7 Polya, 1924, pp. 278–282.
8 Fricke, Klein, 1897.
9 Schönflies, 1891.
10 Fedorov, 1891.

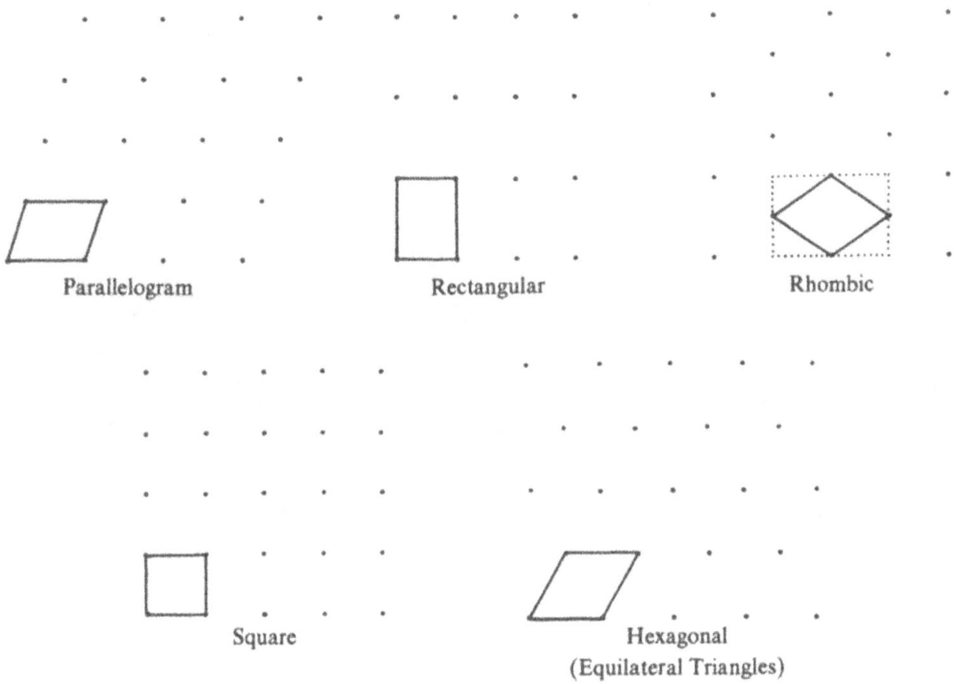

FIGURE 7.20

least possible motive and shows how the motive is repeated. It is not unique, sometimes, you may find different lattices for one ornament. There are five different lattices in two dimensions

In two dimensions there are 6 possible transformations:

1. Translations from the translation group
2. Reflections or 2_a reflections and translations (Gleitspiegelungen) with respect to an axis
3. Rotations of 2nd order (180°) around axes perpendicular to the plane of the ornament
4. Rotations of third order (120°) around axes perpendicular to the plane of the ornament
5. Rotations of fourth order (90°) around axes perpendicular to the plane of the ornament
6. Rotations of sixth order (60°) around axes perpendicular to the plane of the ornament.

Let us now try to find those various patterns in Arabic ornaments. Instead of Polya's table, I shall use the designs given by Andreas Speiser for each of the different patterns. For more expert readers, I shall give the name of the pattern both in Speiser's classification and in the international classification.

1) International p1, Speiser C_1

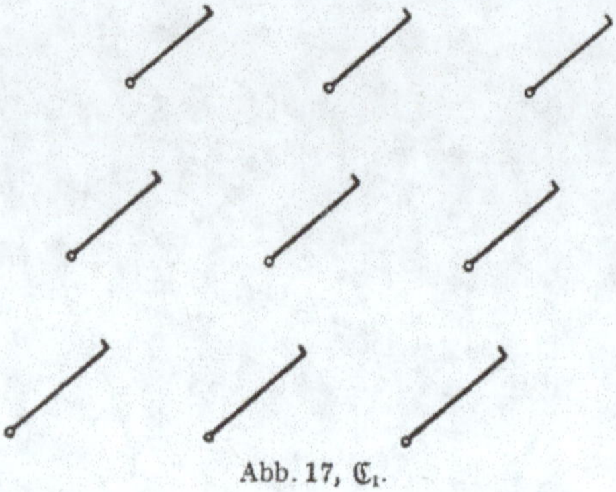

FIGURE 7.21
The motive has no symmetry (mind the little line at the end of the main line). There are only translations in a parallelogrammatic lattice. You could be tempted to see axes of order 3 (120°) because discarding colours gives group International p3 or Speiser C_3.

FIGURE 7.22
Alhambra, azulejos, Salla de Camas (bathroom), Weyl, 1956, p. 113, fig 63
PICTURE FROM THE AUTHOR

2) International p2, Speiser C_2,

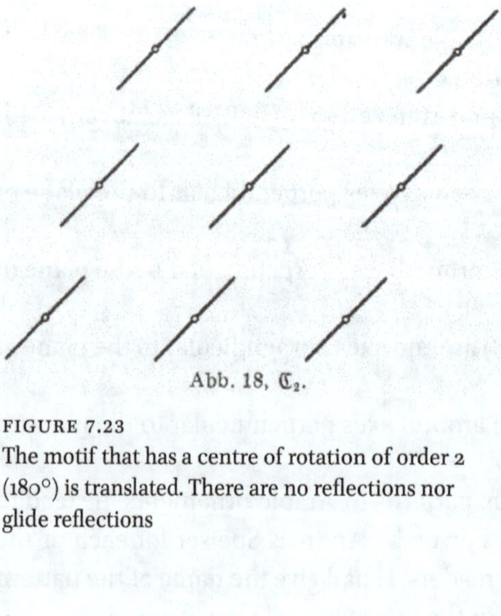

FIGURE 7.23
The motif that has a centre of rotation of order 2 (180°) is translated. There are no reflections nor glide reflections

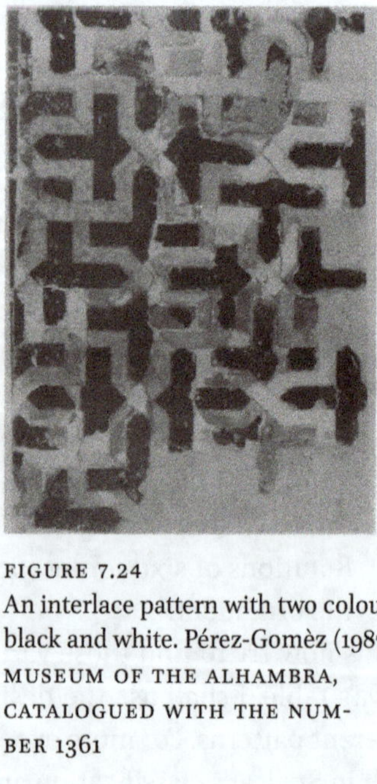

FIGURE 7.24
An interlace pattern with two colours, black and white. Pérez-Gomèz (1987)
MUSEUM OF THE ALHAMBRA, CATALOGUED WITH THE NUMBER 1361

3) International pm, Speiser C^{I}_{s},

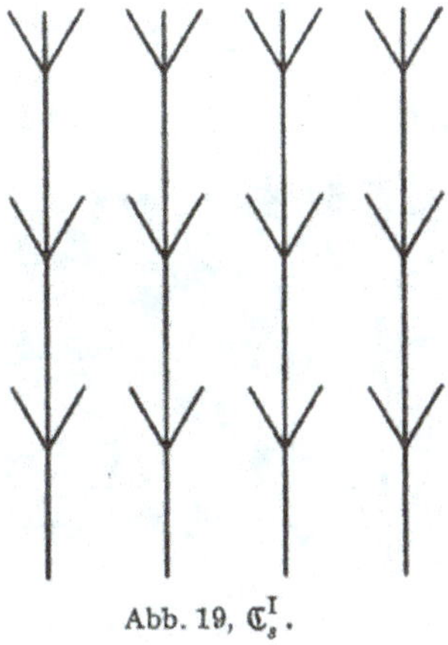

FIGURE 7.25
Lattice is orthogonal and the motive has an axis of reflection. Axes of reflection are parallel to one of the axes of translation and perpendicular to the others.

FIGURE 7.26
Wall tiles of the Kiosk of Mahou Bey (sixteenth century). Prisse d'Avennes (1869–1877), p. 261

4) International pg, Speiser C^{II}_{s}

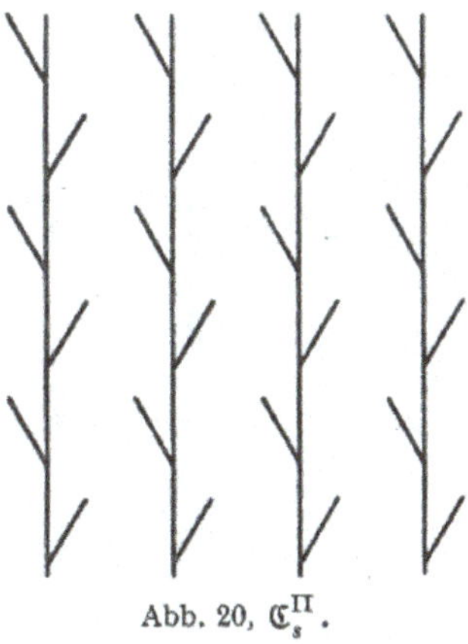

FIGURE 7.27
Lattice is orthogonal. There are axes of glide reflections. Their direction is parallel to the axes of translation and perpendicular to the others. There are neither rotations nor reflections.

FIGURE 7.28
Wall mosaics (seventeenth century). Mosque of El Bordeyny. Prisse d'Avennes (1869–1877), p. 169

5) International cm, Speiser $C^{III}{}_s$

FIGURE 7.29
Lattice is rhomboid. The motive is symmetrical with respect to the axis. Symmetry axes are alternatively axes of reflection and axes of glide-reflection.

FIGURE 7.30
Mural ceramics from a *hanut* (bath), Prisse d'Avennes (1869–1877), p. 297.

6) International pmm, Speiser $C^{I}{}_{2v}$

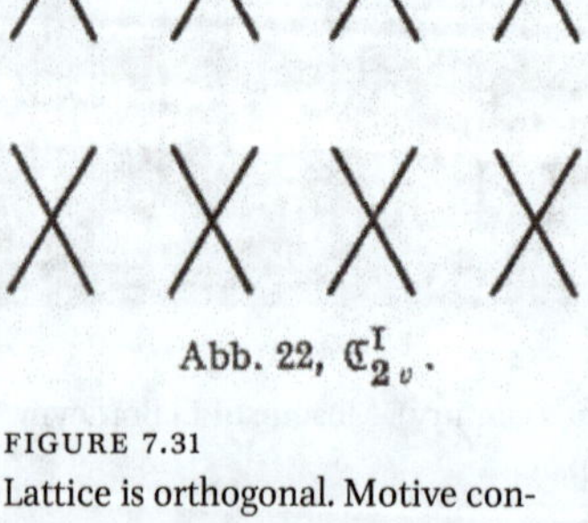

FIGURE 7.31
Lattice is orthogonal. Motive contains a centre of rotation and two perpendicular axes of reflection. All axes of symmetry are reflection axes.

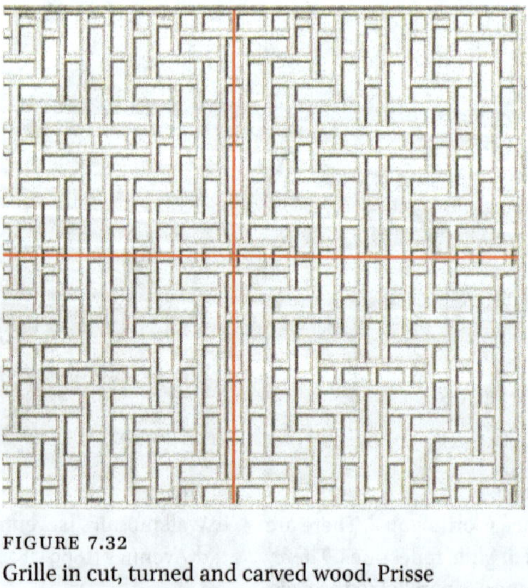

FIGURE 7.32
Grille in cut, turned and carved wood. Prisse d'Avennes (1869–1877), p. 307

7) International pgg, Speiser C^{II}_{2v}

FIGURE 7.33
Lattice is orthogonal. The motive is obtained through glide-reflection from a figure having a centre. All axes of symmetry are glide-reflection axes.

FIGURE 7.34
Floors of several halls (e.g. Hall of the "Aljimeces") and ceilings (e.g. The Door of the Vine), Alhambra. Pérez-Gomèz (1987)

8) International pmg, Speiser C^{3}_{2v}

FIGURE 7.35
Lattice is orthogonal. The motive is obtained through reflection with respect to an axis of a figure having a centre. All axes parallel to that one are reflection axes, axes perpendicular to them are glide-reflection axes.

FIGURE 7.36
Egyptian motive, Thebes' necropole (eighteenth-nineteenth dyn.), Prisse d'Avennes (1879), pl. Architecture; Flinders Petrie (1920), p. 30; Speiser (1956), p. 92.
PHOTO NY PUBLIC LIBRARY

9) International cmm, Speiser C^{IV}_{2v}

FIGURE 7.37
Lattice is rhomboid. Motive has a centre and two axes of reflection passing through it.

FIGURE 7.38
Egyptian motive, Thebes' necropole (eighteenth-nineteenth dyn.), Prisse d'Avennes (1879), pl. Architecture; Flinders Petrie (1920), p. 31; Speiser, 1956, p. 92
PHOTO NY PUBLIC LIBRARY

10) International p4, Speiser C_4

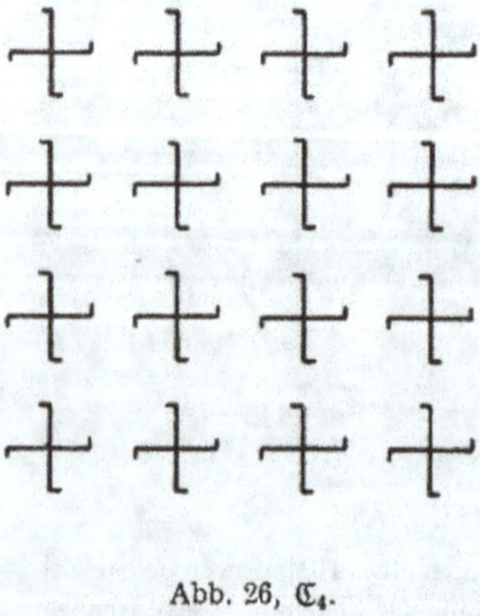

FIGURE 7.39
First group with an axis of rotation of order 4 (90°). It does not contain an axis of reflection. Lattice is square.

FIGURE 7.40
Grille in cut, turned and carved wood. Prisse d'Avennes (1869–1877), p. 307

11) International p4m, Speiser C^{I}_{4v}

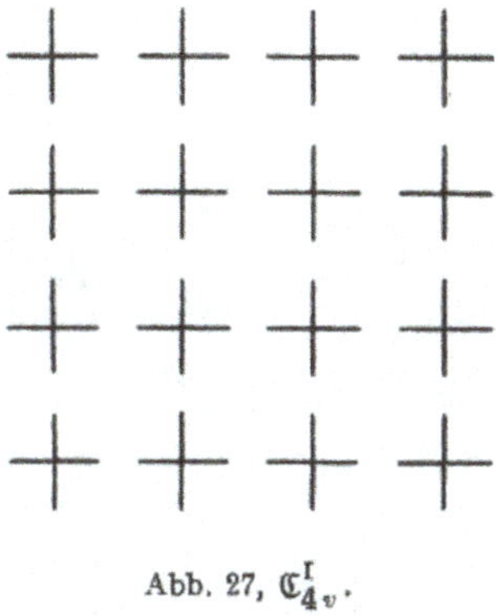

FIGURE 7.41
The lattice is a rhomboid. The ornament has axes of order 4 and four kinds of axes of symmetry, that make angles of 40°. The motive has a symmetry of order four and four axes of reflection passing through its centre of symmetry. The other families of axes are alternatively axes of reflection are axes of glide-reflection.

FIGURE 7.42
Mosque of Qayçun (fourteenth century) Prisse d'Avennes (1869–1877), p. 14

12) International p4g, Speiser C^{II}_{4v}

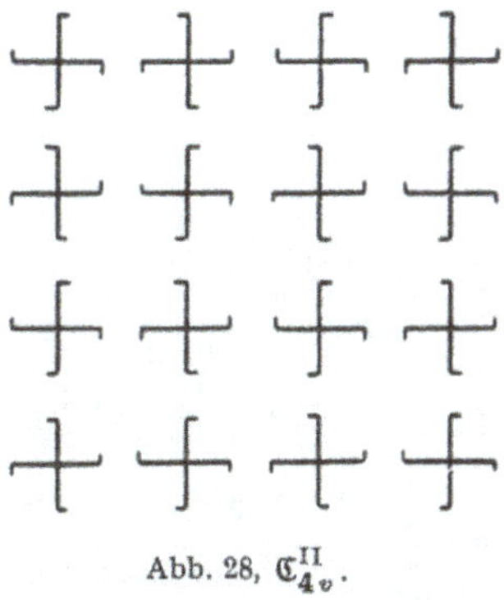

FIGURE 7.43
The ornament has axes of order 4 and four kinds of axes of symmetry, that make angles of 40°. The motive is obtained by reflection of a figure having an axis of order four, by an axis that doesn't pass through the centre. Both families of axes of symmetry parallel to the sides the square pass through the centre and are alternatively reflection and glide-reflection axes. The two other families are all glide-reflection axes.

FIGURE 7.44
Egyptian ornament given by Owen Jones (1856), pl. X, nº 15, by Flinders Petrie, 1920, p. 35, and by Speiser, 1923 (1956), p. 93

13) International p3, Speiser C_3

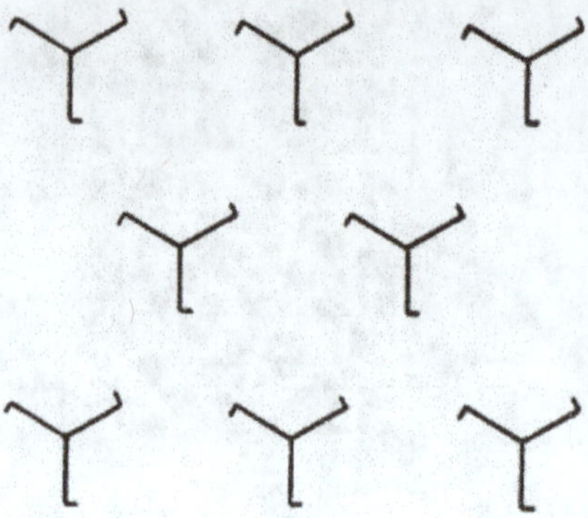

Abb. 29, $\mathfrak{C}_3$.

FIGURE 7.45
This is the simplest group that contains a rotation of order 3 (120°). The motive has an axis of rotation of order three embedded in a hexagonal lattice. The others families are only composed of glide-reflection axes.

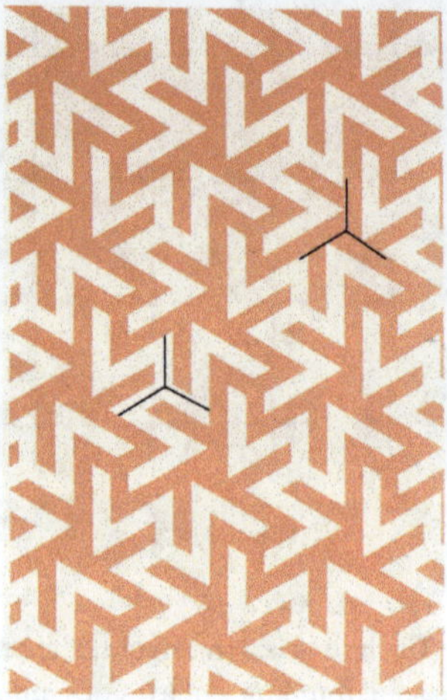

FIGURE 7.46
Stucco inlay on stone (sixteenth-eighteenth century) Prisse d'Avennes (1869–1877), p. 165

14) International p31m, Speiser C^{I}_{3v}

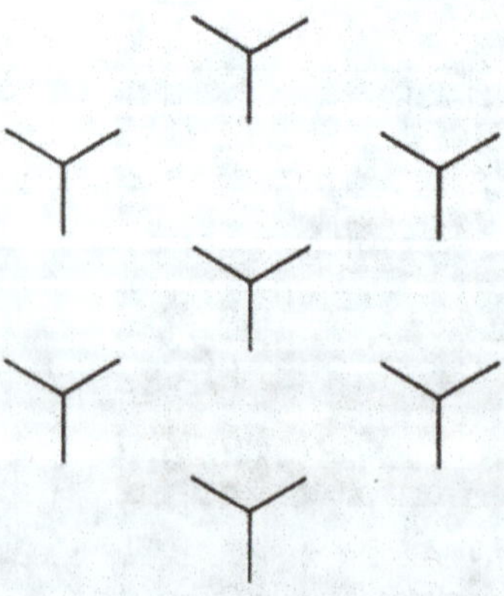

Abb. 30, $\mathfrak{C}^{I}_{3v}$.

FIGURE 7.47
The motive has an axis of rotation of order 3 and three families of axes of symmetry that build an angle of 60 degrees. There are two cases. First, the axes of reflection are parallel to the side of the triangle formed by the reflection axes.

FIGURE 7.48
Ceiling, house called Beyt el-Tchelebey (eighteenth century), Prisse d'Avennes (1869–1877), p. 187 below

15) International p3m1, Speiser C^{II}_{3v}

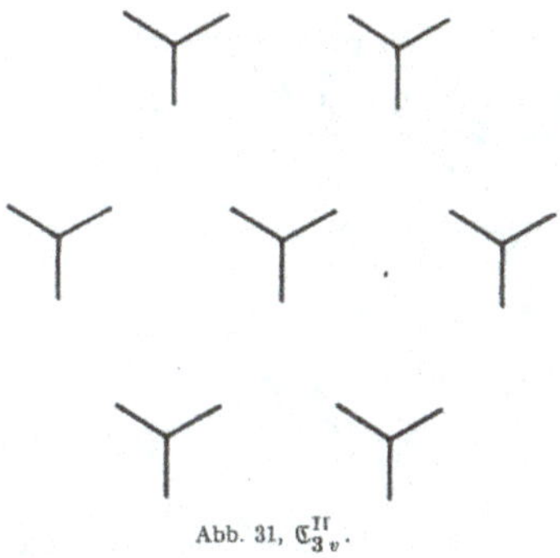

FIGURE 7.49
Second case. The axes of reflection are parallel to the heights of the triangle formed by the reflection axes.

FIGURE 7.50
Madrassa of Amir Sarghatmish in Medieval Cairo (1356)
HTTPS://FOTENT.WORDPRESS.COM/2011/07/05/ARTS-DECORATIVES-DEL-MON-ISLAMIC-GELOSIES/

16) International p6, Speiser C_6

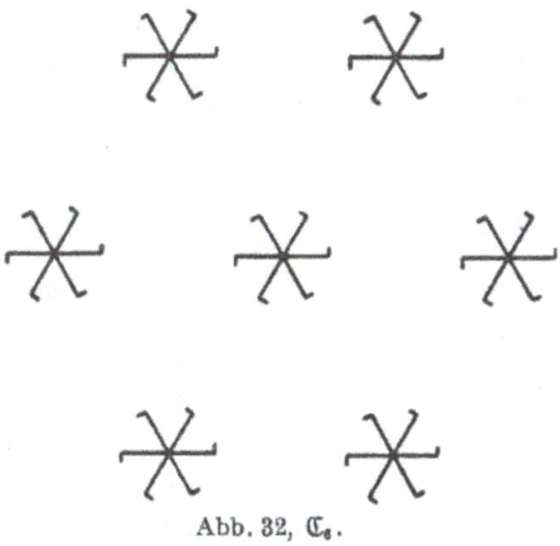

FIGURE 7.51
This group contains an axis of rotation of order 6 (60°) but no reflections. The lattice is hexagonal and also contains rotations of orders 2 and 3, but no reflections.

FIGURE 7.52
Madrassa of Amir Sarghatmish in Medieval Cairo (1356). Prisse d'Avennes (1869–1877), p. 143

17) International p6m, Speiser C_{6v}

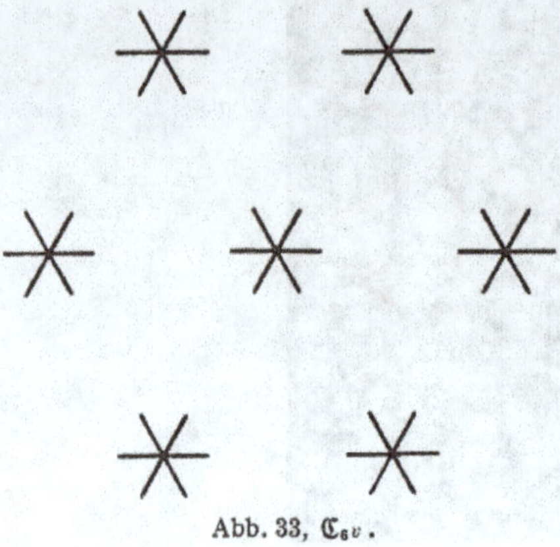

FIGURE 7.53
The ornament has axes of rotation of order 6 and six families of reflection and glide-reflection axis. The ornament has both the symmetries 15 and 16.

FIGURE 7.54
Mosque of Qeyçoun, decoration of the windows inside (fourteenth century). Prisse d'Avennes (1869–1877), p. 143

2 Conclusion

I hope to have convinced you that these designs are not just sketches rapidly done on the corner of a table. They require a deep reflection and we know since the works of Andreas Speiser, Hermann Weyl and some others that that reflection is of a mathematical type. Flinders Petrie, Sir William Matthew wrote in 1920, in his book *Egyptian decorative art*:

> Practically it is very difficult, or almost impossible, to point out decoration which is proved to have originated independently, and not to have been copied from the Egyptian stock.[11]

I hope to have shown how important that research was in the whole Arab world and how important it has been for the development of actual physics. I will say more about that in the second article in this volume.

11 Flinders Petrie, 1920, p. 5, quoted in Speiser, 1956, p. 76.

Bibliography

Fedorov, Evgraf Stepanovich (1853–1919), "Symmetrie der regelmässigen Systeme von Figuren", in *Verhandlungen der Russisch-Kaiserlichen Mineralogischen Gesellschaft zu St. Petersburg* (St. Petersburg, Russia, 1891) 2, 1–145 [first published in Russian in 1891].

Fedorov, Evgraf Stepanovich, "Reguläre Plan und Raumtheilung", *Abh. II Cl. Akad. Wiss. München*, 20 (1900), 465–588.

Flinders Petrie, Sir William Matthew (1853–1942), *Egyptian decorative art* (London: Methuen and Co, 1920).

Fricke, Robert (1861–1930) and Klein, Felix (1849–1925), *Vorlesungen über die Theorie der automorphen Funktionen* (Leipzig: Von B.G. Teubner, 1897).

Grünbaum, B., Grünbaum, Z., and Shepherd, G.C., "Symmetry in Moorish art and other ornaments", *Comput. Maths. Applic.* 12B (1986), 641–653.

Jones, Owen, *Plans, elevations, sections, and details of the Alhambra*, from Drawings taken on the spot in 1834 by Jules Goury, and in 1834 and 1837 by Jones, Owen (London, 1842–1845).

Jones, Owen, (1809–1874), *The grammar of Ornament* (London: Day and son, 1856).

Niggli, Paul, (1888–1953), *Geometrische Krystallographie des Diskontinuums* (Leipzig, 1919).

Pérez-Gòmez, R., "The four regular mosaics missing in the Alhambra", *Comput. Math. Applic.* Vol. 14. N°2 (1987), 33–137

Polya, George, (1887–1985), "Über die Analogie der Krystallsymmetrie in der Ebene", *Zeitschrift für Kristallographie*, 60 (1924), 278–282.

Prisse d'Avennes, Emile, (1807–1879), *L'art arabe d'après les monuments du Kaire, depuis le VIIe siècle jusqu'à la fin du XVIIIe* (Paris: J. Savoie et Cie, 1877) [pagination is taken from the recent Taschen edition].

Prisse d'Avennes, Emile, *Histoire de l'art egyptien, d'après les monuments, depuis les temps les plus reculés jusqu'à la domination romaine*, Texte par P. Marchandon de la Faye (Paris: A. Bertrand, 1879).

Schönflies, Arthur Moritz (1853–1928), *Kristallsysteme und Kristallstruktur* (Leipzig, 1891).

Speiser, Andreas, (1885–1970), Die Theorie der Gruppen von Endlicher Ordnung, mit Anwendungen auf algebraische Zahlen und Gleichungen sowie auf die Krystallographie (Basel: Birkhäuser, 1956 [1st ed. 1923]).

Weyl, Hermann, (1885–1970), *Symmetry* (Princeton: Princeton university Press), 1952.

CHAPTER 8

The Prehistory of the Principle of Relativity

Patricia Radelet-de Grave

Hermann Weyl (1885–1955) whom we already met in the article on Arabic art was one of the greatest mathematicians, physicists and philosophers of the last century. He thought that fundamental ideas were perennial and made this idea into a research method. When tackling a new subject, or taking it up again right from the beginning, he began dissecting it looking for the primordial ideas and then went back in time looking for texts that enunciated similar ideas, or ideas that could lead to those primordial ideas. His aim was to establish a kind of genealogy of ideas and to make an inventory of the various aspects of a particular idea. This method led him to numerous discoveries important for today's mathematics and physics. He shows this in a very clear way in his "Was ist Materie?"

I will try to do the same in the case of relativity and I will follow the development of those ideas until Galilean relativity. Afterwards I will have an *a posteriori* look at that story using texts from Laplace, Weyl and Einstein.

My idea is that fundamental concepts are not only perennial but also universal and it seems to me that if one had enough linguistic knowledge, one could find similar premises of those concepts in other civilisations too. Weyl writes:

> According to the extremely brilliant results which experimental physics in close connection with the theory has obtained in the last decade, the atomic constitution of natural bodies can no longer be doubted. But it is not primarily the structure of the bodies made of indivisible elementary quanta, electrons and atomic nuclei that are to be discussed here. Our question is deeper: What is the "matter" of which these ultimate entities are composed?[1]

1 Nach den überaus glänzenden Ergebnissen, welche die experimentierende Physik in enger Verbindung mit der Theorie in den letzten Dezennien gewonnen hat, kann an der atomistischen Konstitution der Naturkörper kein Zweifel mehr walten. Aber nicht vom Aufbau der Körper aus unteilbaren Elementarquanten, Elektronen und Atomkernen, soll hier in erster Linie die Rede sein, sondern unsere Frage zielt tiefer: was ist die „*Materie*", aus denen diese letzten Einheiten selber bestehen?, Weyl (1924), 28, 561.

 | DOI:10.1163/9789004513402_010

FIGURE 8.1

1 Hermann Weyl Locates His Reflection with Respect to the Knowledge of His Time

At the time, in 1924, one year before Heisenberg discovered the uncertainty principle, physicists knew only about a nucleus in the atom (Ernest Rutherford 1908) and electrons. Weyl goes beyond the constitution of atoms and asks: what is matter itself? Let us see how he starts his reflexion on that argument.

> Since ancient times, Philosophy has tried to answer this question …
>
> But we should try to understand the old philosophical doctrines more precisely from the point of view now obtained in mathematics and physics.
>
> As we are more concerned with the matter itself than with its history; it seems to me justifiable, not to give an objective historical view but a retrospective one from the historical point of view of the present.[2]

---

2 "Seit altersher hat die *Philosophie* darauf eine Antwort zu geben versucht …
Doch soll versucht werden, von dem heute in Mathematik und Physik gewonnenen Standpunkte aus die alten philosophischen Lehren präziser auszudeuten.

Some years before, in 1919, Weyl wrote his famous *Raum, Zeit, Materie* [Space, time, matter] where he continues the work on relativity of Minkowski and of Einstein. His introduction starts with the following words:

> **Space** and **time** are commonly regarded as the forms of existence of the real world, **matter** as its substance. A definite portion of matter occupies a definite part of space at a definite moment of time. It is in the composite idea of **motion** that these three fundamental conceptions enter into intimate relationship. Descartes defined the objective of the exact sciences as consisting in the description of all happening in terms of these three fundamental conceptions, thus referring them to motion.[3]

Weyl enumerates all fundamental quantities as given by Descartes and goes on looking through history:

> Since the human mind first wakened from slumber, and was allowed to give itself free rein, it has never ceased to feel the profoundly mysterious nature of time-consciousness, of the progression of the world in time,—of Becoming. It is one of those ultimate metaphysical problems which philosophy has striven to elucidate and unravel at every stage of its history. The Greeks made Space the subject-matter of a science of supreme simplicity and certainty. Out of it grew, in the mind of classical antiquity, the idea of pure science. Geometry became one of the most powerful expressions of that sovereignty of the intellect that inspired the thought of those times.

Time and space were studied since the beginnings. Time is still a problem and space gave rise to geometry.

> At a later epoch, when the intellectual despotism of the Church, which had been maintained through the Middle Ages, had crumbled, and a wave of scepticism threatened to sweep away all that had seemed most fixed, those who believed in Truth clung to Geometry as to a rock, and it was the highest ideal of every scientist to carry on his science "more geometrico". Matter was imagined to be a substance involved in every change, and it

---

Im übrigen kommt es uns mehr auf die Sache aus auf ihre Geschichte an; um so berechtigter erscheint mir da eine solche nicht objektive, sondern **von dem historischen Augenpunkt der Gegenwart** retrospektive Geschichtsbetrachtung", Weyl (1924) 28, 561.

3 Weyl (1922), 1.

> was thought that every piece of matter could be measured as a quantity, and that its characteristic expression as a "substance" was the Law of Conservation of Matter which asserts that matter remains constant in amount throughout every change.[4]

Weyl underlines the influence of geometry into the development men's mind. The question of matter on the contrary was left a bit aside.

> This, which has hitherto represented our knowledge of space and matter, and which was in many quarters claimed by philosophers as a priori knowledge, absolutely general and necessary, stands today a tottering structure. First, the physicists in the persons of Faraday and Maxwell, proposed the "electromagnetic field" in contra-distinction to matter, as a reality of a different category.[5]

But more recently Faraday and Maxwell proposed a new kind of reality namely "the field".

> Then, during the last century, the mathematician, following a different line of thought, secretly undermined belief in the evidence of Euclidean Geometry. And now, in our time, there has been unloosed a cataclysm which has swept away space, time, and matter hitherto regarded as the firmest pillars of natural science, but only to make place for a view of things of wider scope, and entailing a deeper vision.[6]

Weyl closes his "history" on a revolution, as Kuhn would say. The revolution of general relativity, the object of his book. In this text we shall limit ourselves to special relativity.

I gave that rather lengthy quotation, to show how Weyl could use history in order to find new ideas. This is exactly what Weyl means when he says: "[to get] a retrospective [view] from the historical point of view of the present" and that is what we'll try to do now, starting from the first glim of relativity, namely:

---

4 Weyl (1922), 1.

5 Weyl (1922), 1–2.

6 Weyl (1922), 2.

FIGURE 8.2 Apparent motion of rotation of "fixed" stars around the axis of rotation of the earth

2 Optical Relativity

Let us now follow Weyl's sample and find in the ancient times, the deepest roots of relativity.

My story begins in Antiquity with a well-known text, Virgil's Aeneid, as we shall see it will often be quoted.

After the destruction of Troy Aeneas had to leave. He arrives in Thrace where Polydor let him know that he had been killed by the king of Thrace. When leaving Thrace per ship, Aeneas says:

> Provehimur portu, terraeque urbesque recedunt.[7]

> We sail out of port, and cities and lands recede.

Describing the impression one gets when standing on a ship that "seems at rest" and looking at the coast that one leaves. One doesn't necessarily feel motion, our view may be wrong.

There are various translations, to better understand them, let us give the whole paragraph.

7 Virgilius, Aeneid, III, 72.

FIGURE 8.3 Virgil, 70 BCE–19 BCE

Inde, ubi prima fides pelago, placataque venti
dant maria et lenis crepitans vocat Auster in altum,
deducunt socii navis et litora complent:
provehimur portu, terraeque urbesque recedunt.[8]

The aim of what follows is not to judge the poetical quality of the translations, but to find out translations giving back the relativistic effect. A.S. Kline's translation is correct from our point of view. He writes

Then as soon as we've confidence in the waves, and the winds
grant us calm seas, and the soft whispering breeze calls to the deep,
my companions float the ships and crowd to the shore:
we set out from harbour, and lands and cities recede.

8 Virgilius, Ibid.

On the contrary, John Dryden missed the point. He translates

> Now, when the raging storms no longer reign,
> But southern gales invite us to the main,
> We launch our vessels, with a prosp'rous wind,
> And leave the cities and the shores behind.

French translations are similar. M. De Pongerville writes:

> Nous nous élançons du port, et les champs et les villes s' éloignent.

and Paul Veyne does almost the same

> Nous progressons hors du port, les terres et les villes s' éloignent.

On the contrary l' abbé Delille just as John Dryden missed the point.

> On part, on vole au gré d' un vent rapide et doux;
> Et la ville et le port sont déjà loin de nous.

Some say that Virgil's idea was taken over by William Shakespeare (1564–1616) in *the Tempest* (1611), act II, scene 2. But until now I couldn't find it. Later on it was repeated by Charles Nodier (1780–1844) in *Smarra ou les démonts de la nuit* (1821)

> Pendant ce temps-là les tours, les rues, la ville entière fuyaient derrière moi comme le port abandonné par un vaisseau aventureux qui va tenter les destins de la mer.

We'll find it repeated by Copernic and Laplace, both involved in the formulation of relativity.

FIGURE 8.4 Nicolas Copernicus by Theodore de Bry

3 Nicolas Copernicus (1473–1543)

Copernicus' landmark work *De Revolutionibus orbium celestium* (1543) is based on Virgil's idea. But this text is the last and elaborate expression given by Copernicus. The same idea had already been expressed some 40 years earlier in a manuscript commonly called *Commentariolus* (1508–1514). The idea is expressed right at the beginning in the postulates, because it is the idea that gave birth to Copernicus' work in the *Commentariolus* as well as in *De Revolutionibus* and his whole theory is based on that idea.

> Fifth postulate of the *Commentariolus*.
> Whatever motion appears in the firmament is due, not to it, but to the earth. Accordingly, the earth together with the circumjacent elements performs a complete rotation on its fixed poles in a daily motion, while the firmament and highest heaven abide unchanged.[9]

---

9 "Quinta petitio: Quicquid ex motu apparet in firmamento, non esse ex parte ipsius, sed terrae.

This time, the boat is replaced by the earth. The earth rotates around its axis and the firmament is fixe, its motion is an illusion due to the earth's motion.

> Sixth postulate
> What appear to us as motions of the sun are due, not to its motion, but to the motion of the earth and our sphere, with which we revolve about the sun as [we would with] any other planet. The earth has, then, more than one motion.[10]

Similarly, it is not the sun that rotates around the earth. On the contrary, it the earth that rotates around the sun.

> Seventh postulate
> What appears in the planets as [the alternation of] retrograde and direct motion is due, not to their motion, but to the earth's. The motion of the earth alone, therefore, suffices [to explain] so many apparent irregularities in the heaven.[11]

And the strange motions of the planets, their retrograde motion, are due to the motion of the earth.

2.1 De Revolutionibus (*1543*)

In his major work Copernicus repeats his postulates almost word for word. And he explains more clearly the profound meaning he attributes to Virgil's verse.

> Why then do we still hesitate to grant it [the earth] the motion appropriate by nature to its form rather than attribute a movement to the entire universe, whose limit is unknown and unknowable? Why should we not admit, with regard to the daily rotation, that the appearance is in the heavens and the reality in the earth? This situation closely resembles what

Terra igitur cum proximis elementis motu diurno tota convertitur in polis suis invariabilibus firmamento immobili permanente ac ultimo caelo", Copernicus, *Commentariolus* (1508–1514).

10 "Sexta petitio: Venus nono mense, Mercurius tertio revolutionem peragit. Quicquid nobis ex motibus circa Solem apparet, non esse occasione ipsius, sed telluris et nostri orbis, cum quo circa Solem volvimur ceu aliquo alio sidere, sicque terram pluribus motibus ferri", transl. Ed. Rosen.

11 "Septima petitio: Quod apparet in erraticis retrocessio ac progressus, non esse ex parte ipsarum sed telluris. huius igitur solius motus tot apparentibus in caelo diversitatibus sufficit", transl. Ed. Rosen.

> Vergil's Aeneas says: *Forth from the harbor we sail, and the land and the cities slip backward* [*Aeneid*, III, 72]. For when a ship is floating calmly along, the sailors see its motion **mirrored** in everything outside, while on the other hand they suppose that they are stationary, together with everything on board. In the same way, the motion of the earth can unquestionably produce the impression that the entire universe is rotating.[12]

In this passage, Copernicus explains clearly, the transposition of the idea of the ship to that of the earth and of the Universe.

4 Simon Stevin (1548–1620), *Wisconstige Gedachtenissen* (1585–1608)

Some years after Copernicus, Stevin develops the same idea in his work "reflections". He is one of the first defenders of Copernicus' theory.

In book III, Stevin begins explaining the difficulties encountered by Ptolemy's theory of a fixed earth.

> 6th PROPOSITION.
> To speak of the wondering at what is no wonder, of those who assume a fixed Earth.
>
> Some of those who understand *Ptolemy's* description of the Planetary Motions based on a fixed Earth and consider it correct, are astonished at some properties they perceive therein. Firstly, that Saturn, Jupiter, and Mars, when in opposition to the Sun, always come nearest to the Earth, but when in conjunction, farthest from it. Secondly, that their motion on the epicycle always corresponds exactly to the surplus of the Sun's motion over the motion of the centres of their epicycles. Thirdly, that with Venus and Mercury the converse takes place, for their motion on the epicycle has no such correspondence to the Sun's motion, but the motion of the centre of their epicycle is equal to it. They take this for a sign of the special

12 "Cur ergo hesitamus adhuc, mobilitatem illi formae suae a natura congruentem concedere, magis quam quod totus labatur mundus, cuius finis ignoratur, scirique nequit, necque fateamur ipsius cotidianae revolutionis in caelo apparentiam esse, et in terra veritatem? Et haec perinde se habere, ac si diceret Virgilianus Æneas: Provehimur portu, tarraeque urbesque recedunt. Quoniam fluitante sub tranquilitate navigio, cuncta quae extrinsecus sunt, ad motus illius imaginem moveri cernuntur a navigantibus, ac vicissim se quiescere putant cum omnibus quae secum sunt. Ita nimirum in motu terrae putant cum omnibus quae secum sunt. Ita nimirum in motu terrae potest contingere, ut totus circuire mundus existimetur", Copernicus, *De Revolutionibus* (1543), 6.

WISCONSTIGE
GEDACHTENISSEN,
Inhoudende t'ghene daer hem
in gheoeffent heeft
DEN DOORLVCHTICHSTEN
Hoochgheboren Vorst ende Heere, MAVRITS Prince van
Oraengien, Grave van Nassau, Catzenellenbogen, Vianden, Moers &c.
Marckgraef vander Vere, ende Vlissinghen &c. Heere der Stadt Grave
ende S'landts van Cuyc, St. Vyt, Daesburch &c. Gouverneur van
Gelderlant, Hollant, Zeelant, Westvrieslant, Zutphen
Vtrecht, Overyssel &c. Opperste Veltheer vande
vereenichde Nederlanden, Admirael
generael vander Zee &c.

Beschreven deur SIMON STEVIN *van Brugghe.*

TOT LEYDEN,
Inde Druckerye van Ian Bouwensz.
Int Iaer cIↄ. Iↄ CVIII.

FIGURE 8.5 Simon Stevin's Book III of the Heavenly Motions, of the Finding of the Motions of the Planets by Means of Mathematical Operations, Based on the True Theory of the Moving Earth
Derde Bouck des Hemelloops, van de vinding der dwaelderloopen, deur wisconstighe wercking ghegront op de wesentlicke stelling des roerenden Eertcloots.

character of the worthiest Planet, the Sun, from whose motion the others take their guidance as from a King and move accordingly. But these may be considered erroneous speculations, resulting from the incorrect theory of a fixed Earth. But since this is greatly similar to people who, not being used to sailing, generally ascribe the motion of their ship to other ships,—such as, when they meet them while lying below deck without seeing water or land, they say: how fast that ship outside ours sails, or, if their ship makes a turning, say that the other, which perhaps lies still, moves round theirs—I will use this as an example to illustrate this subject matter.[13]

---

13 "6 Voorstel.

Te segghen vande verwonderinghen sonder wonder der ghene die een vasten Eertcloot stellen.

Ettelicke der gene die Ptolemeus beschrijving der Dwaelderloopen met een vasten Eertcloot verstaen, en voor recht houden, verwonderen hun in sommighe eyghenschappen dieser in mercken: Ten eersten dat Saturnus, Iupiter en Mars in teghenstant der Son

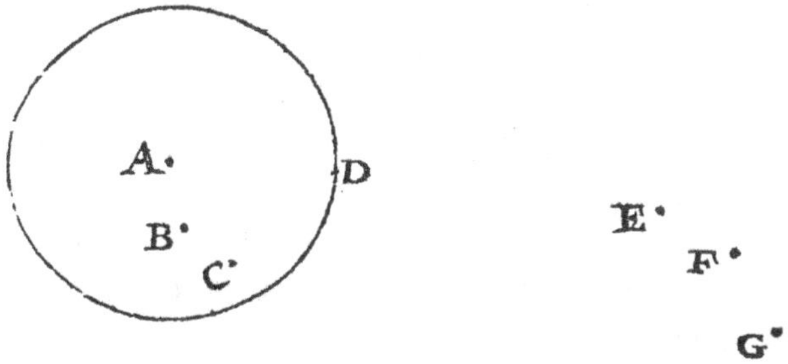

FIGURE 8.6 The Sixth proposition

Then Stevin turns to an example. Taking seven ships, he replaces the sun by ship A and the earth by ship D. Ships B and C are to represent internal planets and E, F, G external ones. He translates then the difficulties encountered by the Ptolemaic system, considering, in terms of a sailor standing on D, the earth. That sailor thinks that his ship is at rest.

> Let these seven points *A, B, C, D, E, F, G* denote seven ships at sea, of which *A*, being the Admiral, lies still. But the ship *D* continually sails in a circle, within which are the three ships *A, B, C*, while the three *E, F, G* are on the outside. This being so, a man being in the ship *D* is of opinion, according to the above, that it lies still and all the others move irregularly about it. And according to this supposition he observes the nature of the motion and, wondering, says as follows: How strange it is that whenever one of the three ships *E, F, G* comes in a straight line from it *via* us to the Admiral *A*, each of them is always nearest to us; and farthest, when it is in this straight line to the other side of the Admiral, however irregular their continual sailing may be! From this he concludes that each of those three ships also

altijt ten naesten by den Eertcloot commen, maer in saming ten versten, Ten tweeden dat haer loop int inront altijt effen overcomt metter overschot des Sonloops boven den loop van haer lnronts middelpunten, Ten derden dat Venus en Mercurius t'verkeerde ghebeurt; want haer loop int inront en heeft mette Son sulcke overcomming niet, maer den loop van haer inronts middelpunt isser even me: Dit houden sy voor een teycken van besonderheyt des weerdichsten Dwaelders de Son, na wiens roersel d'ander als na een Koninck opsicht ncmcn en haer loop vervoughen: Doch men macht houden voor ghedwaelde* spieghelinghen, volghende uyt ghemiste stelling eens vasten Eertcloots. Maer want dit groote ghelijckheyt heeft met luyden die het scheepvaren onghewoon sijnde, ghemeenlick het roersel van haer schip ander schepen toeschrijven, als wanneer sy die teghencommen beneen boort ligghen sondern water of landt de sien, segghen, hoe ras vaert dat schip buyten t'onse. Of hun schip een keer doende, segghen t'ander dat misschien stil light rontom het haer te draeyen, soo sal ick dit als voorbeelt ghebruycken tot verclaring deser stot", Stevin (1605–1608), 259–260, transl E. Crone, E.J. Dijksterhuis, et al. (1961), 139 and 141.

> moves in a smaller circle, in consequence of which they approach and withdraw, wondering why this turning has an exact correspondence in time to the turning of the Admiral. Likewise, considering it strange why the two ships *C, B* also have a connection with the Admiral, but contrary to the foregoing, to wit, that the motion of the large circle they perform in moving round the ship *D* corresponds exactly in time to one turning of the Admiral, he says further that this is a sign of homage paid to the Admiral by the other ships.[14]

But now comes an experienced sailor, who knows that ship D is not at rest but sailing around the Admiral A, just as all other ships B, C, E, F, G. This experienced sailor explains the various wonders the men standing on D had observed.

> This being so, let us assume that a skilled Skipper, knowing what is the matter, said to such a person as follows. You rack your brains in wondering where there is no wonder, for our ship, which you think is lying still, is continually sailing round the three ships *A, B, C*, from which it follows necessarily that whenever we are between the Admiral *A* and one of the three *E, F, G*, each of those must be nearest to us, and farthest from us when *A* is between us and one of them; so that those ships do not sail in circles, with such an imagined motion as makes them approach and withdraw, nor do the two ships *B, C* sail in such circles, corresponding exactly

14 "Laet dese seven punten A, B, C, D, E, F, G, seven schepen in zee beteyckenen waer af A den Admirael sijnde stil light: Maer t'schip D vaert gheduerlick in een rondt, daer de drie schepen A, B, C, binnen sijn, en de drie E, F, G, buyten. Dit soo sijnde, en ymant in het schip D wesende, meynt na de boveschreven ghemeene wijse dattet stille light, en d'ander al rondtom hem onghereghelt draeyen. En volghende sulck gestelde neemt acht op de ghedaente des loops, en seght met een verwonderen aldus: Wat een vreemde saeck ist, dat telckemael als een der drie schepen E, F, G, comt in een rechte lini van hem over ons totten Admirael A, soo is dan elck van dien ons altijt ten naesten: En ten versten, wesende in de selve rechte lini over d'ander sijde vanden Admirael, hoe onghereghelt oock hun gheduerighe vaert is: Hier uyt besluyt hy elck dier drie schepen noch te draeyen in een cleender rondt, daer deur sy naerderen en afwijcken, hem verwonderende waerom sulcken keer in tijt seker overcomming heeft metten keer vanden Admirael: Seghelijcx voor een vreemdicheyt houdende waerom de twee schepen C, B, oock een reghel houden metten Admirael, doch verkeert van de voorgaende, te weten dat den keer des groote ronts diese doen om t'schip D draeyende, in tijt effen overcomt met een keer des Admiraels, seght voort dat sulcx is een teycken van eerbieding die den Admirael van d'ander schepen anghedaen wort", Stevin (1605–1608), 260, transl. 141.

> to the motion of *A*, as you think. But it might be considered unnatural that what seems thus to the inexperienced should not in reality be otherwise.[15]

Now comes the experienced astronomer, who simply translates the discourse of the experienced sailor in terms of astronomy.

> And in the same way an experienced Astronomer might say to an inexperienced one: You rack your brains in wondering where there is no wonder, for our luminary, i.e. the Earth, which you think lies still, travels continually round the three Planets Sun, Venus, Mercury, from which it follows necessarily that whenever we are between the Sun and one of the three: Saturn, Jupiter, Mars, each of them must be nearest to us, and farthest from us when the Sun is between us and one of them; so that those three Planets do not move in circles with such an imagined motion as makes them approach and withdraw, nor do the two, Venus and Mercury, move in such circles, corresponding exactly to the Sun's motion, as you think. But it might be considered unnatural that what seems thus to the inexperienced should not in reality be otherwise.[16]

---

15 "Dit soo sijnde, ghenomen nu dat een ervaren Schipper wetende hoet mette saeck ghestelt is, tegen sulck een aldus seyde: Ghy breeckt u hooft met voor wonder te houden daer geen wonder en is, want ons schip t'welck ghy meent stil te legghen, vaert gheduerlick rondtom de drie schepen A, B, C, waer uyt nootsakelick volght, dat soo dickwils wy sijn tusschen den Admirael A en een der drie E, F, G, soo moet ons dan elck van dien ten naesten sijn, en ten versten als A tusschen ons en een van hemlien is: Inder voughen dat die schepen niet en varen in ronden, met soodanighen versierden loop, die hun doet naerderen en afwijcken, noch oock de twee schepen B, C, in sulcke ronden, efsen overcommende metten loop van A, soo ghy meent: Maer men mochtet voor onnatuerlick houden dat t'ghene voor de onervarenen alsoo schijnt, eyghentlick niet anders en waer", Stevin (1608), 260, transl. 141.

16 "Ende even eens soude een ervaren Hemelmeter tot een onervaren meughen segghen: Ghy breeckt u hooft met voor wonder te houden daer gheen wonder en is, want ons weereltlicht dats den Eertcloot die ghy meent stil te ligghen, draeyt gheduerlick rondtom de drie Dwaelders, Son, Venus, Mercurius, waer uyt nootsakelick volght dat soo dickwils wy sijn tusschen de Son en een der drie Saturnus, Jupiter, Mars, soo moet ons dan elck van dien ten naesten sijn, en ten versten als de Son tusschen ons en een van hemlien is: Inder voughen dat die drie Dwaelders niet en draeyen in ronden met soodanighen versierden loop die hun doet naerderen en afwijcken, noch oock de twee Venus en Mercurius in sulcke ronden, effen overcommende mene Sonloop soo ghy meent, maer men mochtet voor onnatuerlick houden dat t'ghene voor d'onervaernen soo schijnt niet eyghentlick anders en waer", Stevin (1605–1608), 260–261, transl. 141 and 143.

Stevin may now conclude explaining his well-known motto: "Wonder and it is no miracle" that one finds at the head of several of his publications. Stevin explained all he could explain with help of optical relativity. He closes an epoch that one could call the epoch of optical illusions.

> Many people also wonder at the curious jumbling of the motion in latitude of the Planets Saturn, Jupiter, Mars, Venus, and Mercury, based on the theory of a fixed Earth. But according to the theory of a moving Earth there is nothing astonishing, but they are simple orbits deviating from the ecliptic, like the Moon's orbit, from which follow computations of the latitudes with knowledge of the causes, as will appear in its place.
>
> CONCLUSION. We have thus spoken of the wondering at what is no wonder, of those who assume a fixed Earth.[17]

5 Galilean Relativity

Giordano Bruno opens a new epoch when he says that a stone falls at the foot of the mast, when the ship stands still or when it is in motion. For him, it is no more than a simple observation but his idea will be taken up by Galileo in a much larger sense.

Galileo's first idea was to show that earth could be in motion even if we don't feel that motion. And that is the aim of his "Dialogo sopra i due sistemi del mondo" (1632) but this idea will evolve into a much more powerful tool.

On both title pages (8,7), we have the same main characters. Aristotle and Ptolemy on the left and Copernicus on the right. The three of them hold the system of the world they are defending in their hands. Between Ptolemy and Copernicus, we can see the ship that plays such an important role in the *Dialogo*. We all know the protagonists of the *Dialogo*, Sagredo the open minded man, Simplicius the Aristotelian, and Salviati who represents Galileo.

Salviati enumerates arguments against the earth's motion, among them the argument of the stone falling from the mast of a ship, first when at rest and then when in motion:

---

17 "Noch isser by veelen een verwonderen vande seltsaem haspeling des breedeloops der Dwaelders Saturnus, Jupiter, Mars, Venus, en Mercurius, ghegront op stelling eens vasten Eertcloots: Maer volghens de stelling eens roerenden Eertcloots, soo en isser gheen wonder, dan sijn eenvoudelicke weghen afwijckende vanden duysteraer, ghelijck de Maenwech, waer uyt rekeninghen der breede volghen met kennis der oirsaken, als t'sijnder plaets blijcken sal.

FIGURE 8.7 Galileo's *Dialogo*, 1632 and 1635, both published during Galileo's lifetime

SALV. All, for the strongest reason, alledge that of grave bodies, which falling downwards from on high, move by a right line, that is perpendicular to the surface of the Earth, an argument which is held undeniably to prove that the Earth is immoveable: for in case it should have the diurnal motion, a Tower, from the top of which a stone is let fall, being carried along by the conversion of the Earth, in the time that the stone spends in falling, would be transported many hundred yards Eastward, and so far distant from the Towers foot would the stone come to ground. The which effect they back with another experiment; to wit, by letting a bullet of lead fall from the round top of a Ship, that lieth at anchor, and observing the mark it makes where it lights, which they find to be neer the partners [foot] of the Mast; but if the same bullet be let fall from the same place when the ship is under sail, it shall light as far from the former place, as the ship hath run in the time of the leads descent; and this for no other

T'BESLUYT. Wy hebben dan gheseyt vande verwonderinghen sonder wonder der ghene die een vasten Eertcloot stellen.", Stevin (1605–1608), 261, transl. 143.

> reason, than because the natural motion of the ball being at liberty is by a right line towards the centre of the Earth.[18]

One immediately observes that things differ from what was done by Stevin who interpreted the observations as optical illusions. Salviati looks now for observable consequences of the motion of the earth and will show that there are no such consequences. Twenty pages later, he says:

> for whoever shall examine the same, shall find the event succeed quite contrary to what hath been written of it: that is, he shall see the stone fall at all times in the same place of the Ship, whether it stand still, or move with any whatsoever velocity. So that the same holding true in the Earth, as in the Ship, one cannot from the stones falling perpendicularly at the foot of the Tower, conclude any thing touching the motion or rest of the Earth.[19]

Galileo's has thus to convince people of the motion of the earth without any experience able to prove it. That is the situation that will drive him to a much more powerful principle, namely the Galilean principle of relativity or Galilean principle of invariance. Let us read in what a wonderful manner he explains it.

> And here for a final proof of the nullity of all the experiments before alledged, I conceive it now a time and place convenient to demonstrate a

18 "SALV. Per la piú gagliarda ragione si produce da tutti quella de i corpi gravi, che cadendo da alto a basso vengono per una linea retta e perpendicolare alla superficie della Terra; argomento stimato irrefragabile, che la Terra stia immobile: perché, quando ella avesse la conversion diurna, una torre dalla sommità della quale si lasciasse cadere un sasso, venendo portata dalla vertigine della Terra, nel tempo che 'l sasso consuma nel suo cadere, scorrerebbe molte centinaia di braccia verso oriente, e per tanto spazio dovrebbe il sasso percuorere in terra lontano dalla radice della torre. Il quale effetto confermano con un'altra esperienza, cioè col lasciar cadere una palla di piombo dalla cima dell'albero di una nave che stia ferma, notando il segno dove ella batte che è vicino al piè dell'albero; ma se dal medesimo luogo si lascerà cadere la medesima palla quando la nave cammini, la sua percossa sarà lontana dall'altra per tanto spazio quanto la nave sarà scorsa innanzi nel tempo della caduta del piombo, e questo non per altro se non perché il movimento naturale della palla posta in sua libertà è per linea retta verso 'l centro della Terra", Galileo (1632), 118–119, transl. Salusbury (1661), 108.

19 "chiunque la farà, troverà l'esperienza mostrar tutto 'l contrario di quel che viene scritto: cioè mostrerà che la pietra casca sempre nel medesimo luogo della nave, stia ella ferma o muovasi con qualsivoglia velocità. Onde, per esser la medesima ragione della Terra che della nave, dal cader la pietra sempre a perpendicolo al piè della torre non si può inferir nulla del moto o della quiete della Terra", Galileo (1632), 137–138, transl. 126.

way how to make an exact trial of them all. Shut your self up with some friend in the grand Cabbin between the decks of some large Ship, and there procure gnats, flies, and such other small winged creatures: get also a great tub (or other vessel) full of water, and within it put certain fishes; let also a certain bottle be hung up, which drop by drop letteth forth its water into another bottle placed underneath, having a narrow neck: and, the Ship lying still, observe diligently how those small winged animals fly with like velocity towards all parts of the Cabin; how the fishes swim indifferently towards all sides; and how the distilling drops all fall into the bottle placed underneath. And casting any thing towards your friend, you need not throw it with more force one way then another, provided the distances be equal: and leaping, as the saying is, with your feet closed, you will reach as far one way as another. Having observed all these particulars, though no man doubteth that so long as the vessel stands still, they ought to succeed in this manner; make the Ship to move with what velocity you please; for (so long as the motion is uniform, and not fluctuating this way and that way) you shall not discern any the least alteration in all the forenamed effects; nor can you gather by any of them whether the Ship doth move or stand still.

In leaping you shall reach as far upon the floor, as before; nor for that the Ship moveth shall you make a greater leap towards the poop than towards the prow; howbeit in the time that you staid in the Air, the floor under your feet shall have run the contrary way to that of your jump; and throwing any thing to your companion you shall not need to cast it with more strength that it may reach him, if he shall be towards the prow, and you towards the poop, then if you stood in a contrary situation; the drops shall all distill as before into the inferiour bottle and not so much as one shall fall towards the poop, albeit whil'st the drop is in the Air, the Ship shall have run many feet; the Fishes in their water shall not swim with more trouble towards the fore-part, than towards the hinder part of the tub; but shall with equal velocity make to the bait placed on any side of the tub; and lastly, the flies and gnats shall continue their flight indifferently towards all parts; nor shall they ever happen to be driven together towards the side of the Cabbin next the prow, as if they were wearied with following the swift course of the Ship, from which through their suspension in the Air, they had been long separated; and if burning a few graines of incense you make a little smoke, you shall see it ascend on high, and there in manner of a cloud suspend it self, and move indifferently, not inclining more to one side than another: and of this correspondence of effects the cause is for that the Ships motion is common to all the things contained in it, and to the Air also; I mean if those things be shut up in the Cabbin: but in case those things were above deck in the open Air, and not obliged to follow the course of the Ship, differences more or lesse notable would be observed in some of the fore-named effects, and there is no doubt but that the smoke would stay behind as much as the Air it self; the flies also, and the gnats being hindered by the Air would not be able to follow the motion of the Ship, if they were separated at any distance from it. But keeping neer thereto, because the Ship itself as being an unfractuous Fabrick, carrieth along with it part of its neerest Air, they would follow the said Ship without any pains or difficulty. And for the like reason we see sometimes in riding post, that the troublesome flies and the hornets do follow the horses flying sometimes to one, sometimes to another part of the body, but in the falling drops the difference would be very small; and in the salts, and projections of grave bodies altogether imperceptible.[20]

---

20 "E qui, per ultimo sigillo della nullità di tutte le esperienze addotte, mi par tempo e luogo di mostrar il modo di sperimentarle tutte facilissimamente. Riserratevi con qualche amico nella maggiore stanza che sia sotto coverta di alcun gran navilio, e quivi fate d'aver mosche, farfalle e simili animaletti volanti; siavi anco un gran vaso d'acqua, e dentrovi de' pescetti; sospendasi anco in alto qualche secchiello, che a goccia a goccia vadia versando dell'acqua in un altro vaso di angusta bocca, che sia posto a basso: e stando ferma la nave, osservate diligentemente come quelli animaletti volanti con pari velocità vanno verso tutte le parti della stanza; i pesci si vedranno andar notando indifferentemente per tutti i versi; le stille cadenti entreranno tutte nel vaso sottoposto; e voi, gettando all'amico alcuna cosa, non piú gagliardamente la dovrete gettare verso quella parte che verso questa, quando le lontananze sieno eguali; e saltando voi, come si dice, a piè giunti, eguali spazii passerete verso tutte le parti. Osservate che avrete diligentemente tutte queste cose, benché niun dubbio ci sia che mentre il vassello sta fermo non debbano succeder cosí, fate muover la nave con quanta si voglia velocità; ché (pur che il moto sia uniforme e non fluttuante in qua e in là) voi non riconoscerete una minima mutazione in tutti li nominati effetti, né da alcuno di quelli potrete comprender se la nave cammina o pure sta ferma: voi saltando passerete nel tavolato i medesimi spazii che prima né, perché la nave si muova velocissimamente, farete maggior salti verso la poppa che verso la prua, benché, nel tempo che voi state in aria, il tavolato sottopostovi scorra verso la parte contraria al vostro salto; e gettando alcuna cosa al compagno, non con piú forza bisognerà tirarla, per arrivarlo, se egli sarà verso la prua e voi verso poppa, che se voi fuste situati per l'opposito; le gocciole cadranno come prima nel vaso inferiore, senza caderne pur una verso poppa, benché, mentre la gocciola è per aria, la nave scorra molti palmi; i pesci nella lor acqua non con piú fatica noteranno verso la precedente che verso la sussequente parte del vaso, ma con pari agevolezza verranno al cibo posto su qualsivoglia luogo dell'orlo del vaso; e finalmente le farfalle e le mosche continueranno i lor voli indifferentemente verso tutte le parti, né mai accaderà che si riduchino verso la parete che riguarda la poppa, quasi che fussero stracche in tener dietro al veloce corso della nave, dalla quale per lungo tempo, trattenendosi per aria, saranno state separate; e se abbruciando alcuna lagrima d'incenso si farà un

FIGURE 8.8 Gregorio a Sancto Vincenzo and the illustration of one of his theses

This vignette could be an illustration of Galileo's text. Nevertheless, it was published some years before by a Jesuit, Gregorio a Sancto Vincenzo. The booklet containing his theses that had to be defended by two of his students was published at Louvain in 1624.

poco di fumo, vedrassi ascender in alto ed a guisa di nugoletta trattenervisi, e indifferentemente muoversi non piú verso questa che quella parte. E di tutta questa corrispondenza d'effetti ne è cagione l'esser il moto della nave comune a tutte le cose contenute in essa ed all'aria ancora, che per ciò dissi io che si stesse sotto coverta; ché quando si stesse di sopra e nell'aria aperta e non seguace del corso della nave, differenze piú e men notabili si vedrebbero in alcuni de gli effetti nominati: e non è dubbio che il fumo resterebbe in dietro, quanto l'aria stessa; le mosche parimente e le farfalle, impedite dall'aria, non potrebber seguir il moto della nave, quando da essa per spazio assai notabile si separassero; ma trattenendovisi vicine, perché la nave stessa, come di fabbrica anfrattuosa, porta seco parte dell'aria sua prossima, senza intoppo o fatica seguirebbon la nave, e per simil cagione veggiamo tal volta, nel correr la posta, le mosche importune e i tafani seguir i cavalli, volandogli ora in questa ed ora in quella parte del corpo; ma nelle gocciole cadenti pochissima sarebbe la differenza, e ne i salti e ne i proietti gravi, del tutto impercettibile", Galileo (1632), 180–182, transl., 165–167.

THEOREMATA MATHEMATICA
SCIENTIÆ STATICÆ.
DE DVCTV PONDERVM
PER PLANITIEM
Rectà & obliquè Horizontem decussantem.
DEFENDENDA AC DEMONSTRANDA
In Collegio Societatis IESV Louanij
A GVALTERO VAN AELST ANTVERPIENSI
PRÆSIDE
R. P. GREGORIO A S. VINCENTIO
Math. Profess. eiusdem Societatis Religiosis,
Die 29. Iulij ante Meridiem, Anno 1624.

FIGURE 8.9 The title page of Gregorio a Sancto Vincenzo's publication

6 Christiaan Huygens

Huygens will be the first to use Galilean invariance in order to find new laws of nature. He will find that way the laws of collision of two bodies. He takes as an axiom Descartes' first law of collision that says that two equal bodies that collide directly with equal velocities will both reverse their motion and come out with equal but opposite velocities. He always speaks of uniform velocities.

Then he will perform the same experiment on a ship moving with respect to the riverside with the same velocity as one of the bodies and thus with a velocity inverse from the other body.

What does the man on the riverside see? He sees one body at rest and the other that arrives with twice the velocity it has on the ship. What happens on the ship after the bodies collide? Descartes' axiom says that they return both with the same velocity but in opposite direction. But what does the man on the bank see? He sees the entering body standing still and the other one going back with twice the velocity it had on the ship.

CHRISTIANUS HUGENIUS

DE

MOTU CORPORUM

EX

PERCUSSIONE.

FIGURE 8.10 Title page of Huygens manuscript "De motu Corporum ex Percussione" [The Motion of Colliding Bodies]

Huygens has found a new law, when a body with a certain velocity, collides with another equal body at rest, it then stands still and the other one will move with the velocity that the first one had before they collided.

Huygens proceeds using always the same principle, namely Galilean invariance: the laws of nature are the same on the ship as on the riverside, if the ship, the system of reference, has a uniform velocity. Proceeding like that, Huygens discovers all the laws of collision between two spherical bodies of equal mass. Adding the principle of conservation of energy, he also finds the laws for unequal bodies.

7 Pierre-Simon Laplace and Hermann Weyl as a Conclusion

Laplace starts his famous book, *The system of the world* with two chapters having titles that remind the procedure used by Stevin and by Galileo as well. The title of the first book is "Of the **apparent** motions of the heavenly bodies" and right under the title of the second book "Of the **real** motions of the heavenly bodies", Laplace reminds us of Virgil's sentence *Provehimur portû, terræque urbesque recedunt*. In doing so Laplace establishes a link between the simple idea expressed by Virgil and his discovery of the motions of heavenly bodies. He does that for good reasons, as he explains it in a kind of preamble to that chapter.

> If man had confined himself to collecting facts alone, science would only have presented a sterile nomenclature, and he would never have attained the knowledge of the great laws of nature. It is by comparing phenomena together, and by endeavouring to trace their connection with each other, that he has succeeded in discovering these laws, the existence of which may be perceived even in the most complicated of their effects. Then it is, that nature in discovering herself, has shown how from a small number of general causes she has produced the infinite variety of phenomena which have been observed; and thus enabled us to determine those, which successive circumstances will bring to light, and being assured that nothing will interpose between these causes and their effects, we venture to extend our views into futurity, and contemplate the series of events, which in time alone can develop. It is only in the theory of the system of the world, that the human understanding has attained to this state of perfection.[21]

Laplace is right, Copernicus, Kepler and Laplace himself adopted the heliocentric system of the world because it was simpler than Ptolemy's old one. History shows that that idea of simplicity is also essential. And that is the reason why Weyl is right when going back in history looking for simple but essential ideas. Virgil's idea will drive him to invariance, an idea derived from the symmetry Arabic art had developed so intensely.

---

21 "Si l'homme s'étoit borné à recueillir des faits; les sciences ne seroient qu'une nomenclature stérile, et jamais il n'eût connu les grandes loix de la nature. C'est en comparant entr'eux les phénoménes, en cherchant à saisir leurs rapports; qu'il est parvenu à découvrir ces loix toujours empreintes dans leurs effets les plus variés. Alors, la nature en se dévoilant, lui a présenté le spectacle d'un petit nombre de causes générales donnant naissance à la foule des phénomènes qu'il avoit observés; il a pu déterminer ceux que les circonstances successives doivent faire éclore, et lorsqu'il s'est assuré que rien ne trouble l'enchaînement de ces causes à leurs effets, il a porté ses regards dans l'avenir, et la série des événemens que le temps doit développer, s'est offerte à sa vue. C'est uniquement encore dans la théorie du systême du monde, que l'esprit humain, par une longue suite d'efforts heureux, s'est élevé à cette hauteur. Essayons de tracer la route la plus directe pour y parvenir", Laplace, *Exposition du système du monde* (1798) [posthume], transl. Pond (1809).

After the discovery of the fact that the speed of light was a limit velocity, Woldemar Voigt (1850–1919), and Hendrik Antoon Lorentz (1853–928) introduced invariance under new transformations, the Lorentz' transformations. This gave birth to "special relativity" that uses new transformations but is based on the same idea of invariance.

Albert Einstein (1875–1955) and Henri Poincaré (1854–1912) will do one more step in generalizing, giving birth to general relativity. As Einstein puts it

> *The laws of physics must be such that they may be applied with respect to any moving frames of reference.*[22]

Hermann Weyl explained the same thing a bit differently:

> Man introduces the frame into nature so nature must remain the same in whatever frame they are expressed.[23]

They show that the laws of nature are invariant under whatever movement of the reference system. Thus those laws are invariant under much more general transformations. Hermann Weyl will then try to go in the same direction. He thought first that the laws of nature should be invariant under a change of the unit of measure of the system of reference. He thinks of a scale invariance. Galileo had also proposed something of the same nature a long time ago but they both had to reject it.

Nevertheless, this idea evolved in Weyl's mind and became one of the most powerful tools of modern physics, namely Gauge invariance. And he said rightfully

> The important thing is not relativity, moreover it disappears, since the laws of nature must remain the same regardless of the reference frame in which they are described.
>
> The important thing is invariance.[24]

In my first article,[25] I showed invariance with the help of Arabic art and in my second article with the help of the prehistory of a theory erroneously called

---

22 *Die Gesetze der Physik müssen so beschaffen sein, dass sie in bezug auf beliebig bewegte Bezugssysteme gelten*, Einstein (1916).

23 Weyl (1948) (1949).

24 Weyl (1948) (1949).

25 "Art and Mathematics, two Different Paths to the Same Truth", in this volume.

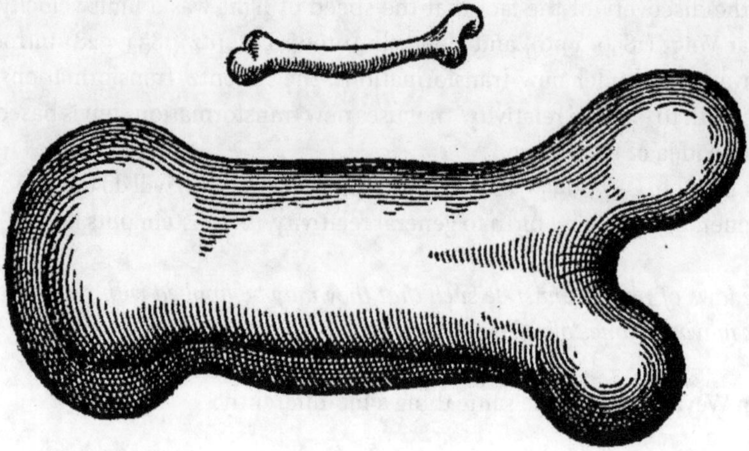

FIGURE 8.11 The idea of scale invariance by Galileo, *Discorsi*, 129

relativity. There is no relativity in Arabic art but there is invariance just as in Virgil's verse. Couldn't we say, to conclude, that modern physics is a product of the wedding of Virgil's verse with the symmetries of Arabic art?

Bibliography

Copernicus, Nicolaus, *Commentariolus* (1508–1514), http://www.fh-augsburg.de/%7Eharsch/Chronologia/Lspost16/Copernicus/kop_c00.html

(URL accessed 14 April 2021), transl. Rosen, Edward, *Three Copernican Treatises: The Commentariolus of Copernicus; The Letter against Werner; The Narratio Prima of Rheticus* (Second revised Edition New York, NY: Dover Publications, Inc., 2004 [1939]).

Copernicus, Nicolaus, *De Revolutionibus* (Nürenberg: Preteius, 1543), transl. Rosen, Edward, http://www.webexhibits.org/calendars/year-text-Copernicus.html (URL accessed 14 2021).

Einstein, Albert, "Die Grundlage der allgemeinen Relativitätstheorie", *Annalen der Physik*, Band 354, Nr. 7, 1916, 769–782.

Galilei, Galileo, *Dialogo sopra i due sistemi del mondo* (Florence: Landini, 1632), transl. by Thomas Salusbury, *Dialogues on two world systems* (London: Leybourne, 1661).

Huygens, Christiaan, "De motu corporum ex percussione", in *Les Œuvres complètes de Christiaan Huygens*, vol. 17 (La Haye: Martin Nijhof, 1929), 29–168, original and translation [posthume].

Laplace, Pierre Simon, *Exposition du système du monde* (Paris: Duprat, 1798) [posthume], transl. Pond, J., *The system of the world* (London: Phillips, 1809).

Stevin, Simon, *Wisconstighe Ghedachtenissen* [1605–1608] (Leiden: Ian Bouwensz,

1608), transl. E. Crone, E.J. Dijksterhuis, et al. in *the Principal works of Simon Stevin* (Amsterdam: Swets & Zeitlinger, 1961).

Weyl, Hermann, *Raum, Zeit, Materie* (Berlin: Springer, 1919), transl. Brose, Henry L., *Space, time, matter* (London: Methuen and Co. 1922),

Weyl, Hermann, "Was ist Materie?", *Die Naturwissenschaften*, Heft 28, 11 Juli 1924, 561–568; Heft 29, 18 Juli 1924, 586–593; Heft 30, 25 Juli 1924, 604–611.

Weyl, Hermann, "Wissenschaft als symbolische Konstruktion des Menschen", *Eranos-Jahrbuch 1948*, (1949), 375–431, Gesammelte Abhandlungen, Bd. 4, 289–346.

CHAPTER 9

Intersections between Social and Scientific Thought: The notion of *muṭābaqa* in the *Muqaddima* of Ibn Khaldūn

Giovanna Lelli

فإنا إذا نظرنا اصل الشيء و جنسه و فصله و مقدار عظمه و قوته أجرينا الحكم في نسبة ذلك على أحواله و حكمنا بالامتناع على ما خرج من نطاقه[1]

> When we carefully consider the essence of a thing, its genus, property, size and power, we can judge its conditions on those grounds and conclude that all that falls outside this domain is impossible.[2]
>
> IBN KHALDŪN, *The Muqaddima*

Scientific thought, like all other aspects of human thought, developed thanks to exchanges and intersections between different disciplines and fields. The European philosophical thought of the seventeenth and eighteenth centuries has been particularly symptomatic of this phenomenon. At that time the diffusion of materialism in the sciences of nature, society and human psychology reflected the refutation of metaphysics as an explanation of worldly phenomena. René Descartes (1596–1650), in his physics, formulated a purely mechanistic conception of matter ("*res extensa*");[3] Thomas Hobbes (1588–1679)

1 Ibn Khaldūn, *Muqaddima*, ed. Quatremère, *Prolégomènes d'Ebn Khaldoun*, volumes I–III (1858. Reprint 1970), I, 329.

2 All translations are the author's, unless otherwise noted. The title of his entire work is *Kitāb al-ʿIbar wa dīwān al-mubtadaʾ wa al-khabar fī ayyām al-ʿArab wa al-ʿAjam wa al-Barbar wa man ʿāṣarahum min dhawī al-sulṭān al-akbar* (*Book of Lessons and Archive of Early and Subsequent History, Dealing with the Political Events Concerning the Arabs, non-Arabs, and Berbers, and the Supreme Rulers Who Were Contemporary with Them*); for a more literal translation cfr. Ibn Khaldūn, (1980), I, 371–372.

3 Descartes, *Principia Philosophiae* (1644).

 | DOI:10.1163/9789004513402_011

provided a mechanistic description of reality, human reason and passions;[4] John Locke (1632–1704) considered experience as the only source of human ideas and refuted all sort of apriorism;[5] David Hume (1711–1776) wanted to apply the method of modern physics for explaining the functioning of human intellect.[6] It was a period of transition. These intersections led to a distinction between the field of the sciences of nature and the field of the social sciences. A decisive contribution in this respect has been given by Isaac Newton's (1642–1727) physics and Karl Marx's (1818–1883) materialistic conception of history.

The purpose of this article is to show that the materialistic views of the Arab historian Ibn Khaldūn (1332–1406) expressed in his book known as the *Muqaddima* ("Introduction"), although geographically and chronologically far from seventeenth- and eighteenth-century Europe, anticipated similar intersections between materialism of nature and materialism of society. In particular, we intend to analyse the meaning of a key-notion in Ibn Khaldūn's historiography: the notion of "*muṭābaqa*". This word literally means "coincidence", "correspondence", "conformity" between superimposable entities. In the field of historiography Ibn Khaldūn uses this word with the meaning of coincidence between historical events (*waqāʾiʿ*) and conditions or circumstances (*aḥwāl*). According to Ibn Khaldūn, in history there is necessarily "coincidence" between events and conditions. *Muṭābaqa* is at once a universal objective law of history, the instrument at the disposal of the historian for verifying the truthfulness of historical sources, and the proof of the veracity of his deductions when they "coincide" with reality.

The *Muqaddima* includes the real introduction and the first book of the work in three books by Ibn Khaldūn, the *Book of Lessons*.[7] The entire book deals with the history of the Arabs and the Berbers, while the book known as *Muqaddima* deals mainly with the views of his author on historiography (introduction, chapters one to four), and the Arab-Islamic arts and sciences (chapters five to six). Since its first French translation, which was the first translation into a Western European language (1862–1868),[8] the historiographic views exposed by Ibn Khaldūn have aroused the admiration of scholars for their scientific, rationalist and empiric approach, which remained without equals for centuries. The *Muqaddima* is not only a book on Ibn Khaldūn's philosophy of history.

---

4 Hobbes, *Leviathan* (1651).

5 Locke, *An Essay Concerning Human Understanding Understanding* (1689).

6 Hume, *A Treatise of Human Nature* (1739–1740).

7 Ibn Khaldūn (1980), I, 13.

8 Ibn Khaldūn, *Muqaddima*, transl. de Slane (1862, 1865 and 1868), quoted in Ibn Khaldūn (1980), I, cviii.

It also a unique synthesis of the Arab culture in the classical era. It is a book of intersections. In this article we analyse the notion of *muṭābaqa* as a microcosm of intersections between two great fields: the "system" of classical Arab culture and Khaldūn's new materialistic conception of history. The latter, in its turn, lies in a zone of intersection between the natural and the social sciences.

We argue that despite Khaldūn's materialism and refutation of metaphysics, the law of *muṭābaqa* remains affected by an *aporia* (a contradiction) consisting in finalism and apriorism. We have identified the following reasons for this aporia in Ibn Khaldūn's philosophy of history:

a. The system of Ibn Khaldūn does not contemplate the impact of human choice on history as a factor that interacts with "conditions".
b. Ibn Khaldūn ascribes to the law of *muṭābaqa* a precision impossible to attain in a social science. Only modern physics will be able to formulate general laws of nature expressed with mathematical formula and experimentally verifiable.
c. Ibn Khaldūn relies upon Avicenna (980–1037)'s *Psychology*, a branch of Aristotelian physics that, since Aristotle himself (384–322 BCE), remained affected by Platonism.

This article is divided into seven sections. In section one, we expose the origin of the Arab root "ṭ b q" according to the first known Arab dictionary, the *Kitāb al-ʿAyn* ("The Book of the Letter ʿAyn") by Khalīl Ibn Aḥmad (721–791).[9] In sections two to five we analyse different occurrences and meanings of the term *muṭābaqa* in the *Muqaddima*. Particularly we analyse its meaning in the field of historiography (section two), in the field of occult sciences and divination (section three), in the field of Islamic philosophy (section four) and in the field of Arab linguistics and logic (section five). In section six we draw the attention of the reader to parallel developments of the root "ṭ b q" in Arab Euclidean geometry. In section seven we analyse the meaning of "nature of civilisation" (*ṭabīʿat al-ʿumrān*) in Ibn Khaldūn's historiography, pointing out that the concept of "nature" in this context lies between Aristotelian physics and modern materialism. In our conclusion we highlight that the finalistic and aprioristic aporia inherent in the historical law of *muṭābaqa* is a most fertile and creative element in Khaldūn's philosophy of history.

9 Throughout this article, the dates of births and deaths of historical figures are often approximate when referred to pre-modern times.

1 The Meaning of the Arab Root "*ṭ b q*" in the *Kitāb al-ʿAyn* by Khalīl Ibn Aḥmad (d. 791)

The *Kitāb al ʿAyn* by Khalīl Ibn Aḥmad al-Farāhīdī[10] is the oldest known Arab dictionary. It mentions meanings and uses of the Arab root "ṭ b q" that help us understanding the meaning of *muṭābaqa* in the *Muqaddima*. We have selected below the meanings that are relevant for our purpose: two milestones of a mill "coincide" (*aṭbaqa*); the "cover" (*ṭabaq*) of a box; "layers" (*ṭabaqāt*) of the sky; "classes" (*ṭabaqāt*) of people; "to agree" (*aṭbaqa*) in opinion; "to conform" (*ṭābaqa*) with somebody's will.

أطبقَ الرحَيين

Two millstones of a mill coincide

الطبق: كل غطاء لازم، ويقال: أطبق الحُقّة ...

The "*ṭabq*" is the appropriate "cover". For example: "He/she has covered" the box [with its lid]

و السموات طبقات بعضها فوق بعض ...

And the skies are "layers" superimposed on one another ...

و الطبق: جماعة من الناس يعدلون طبقةً مثل جماعة ...

"*Ṭabq*" is a group of people who belong to the same "class" (*ṭabaqa*), like a group ...

و أطبق القوم على هذا الأمر أي اجتمعوا و صار كلمتهم واحدةً.

People "coincide" on a matter means that they have the same opinion on it.

و طابقت المرأة زوجها إذا واتته على كل الأمور ...

A woman "conforms" with her husband ['s will] when she satisfies him in all matters ...[11]

10 Khalīl Ibn Aḥmad, *Kitāb al-ʿAyn* (1986).

11 Khalīl Ibn Aḥmad [b], *Kitāb al-ʿAyn*, v (n.d.), 108.

These meanings of the root "ṭ b q" are all helpful to understand the occurrences of *muṭābaqa* in the *Muqaddima* better. Among them, the technical meaning of coincidence between the two stones of a mill is particularly suitable to highlight the meaning of *muṭābaqa* in Ibn Khaldūn's historiography. Just like an engineer and a geometer, the historian measures with the highest accuracy events and their conditions in order to verify their perfect coincidence.

2 Occurrences of *muṭābaqa* in the *Muqaddima*

2.1 *Muṭābaqa* in Ibn Khaldūn's Historiography

The *Muqaddima* is not only a book on Khaldūn's "philosophy of history",[12] but also an inclusive picture of the Arab classical system of thought. All systems contain aporias (contradictions) that undermine them. In this section we analyse the meaning of *muṭābaqa* in Khaldūn's historiography and we highlight the aprioristic aporia that affects it.

We propose to distinguish between two levels in Ibn Khaldūn's philosophy of history. The first level is at once axiomatic and technical. Ibn Khaldūn aims to find *the* universal law of history and *the* methodological instrument that guarantees to the historian an almost geometrical precision of measurements and certitude of results. The second level, although also theoretical, is more concrete and describes a general development model to which all societies conform. The first level applies to micro dimensions and *courte durée* phenomena, namely single events. The second level applies to macro dimensions and *longue durée* processes of about one century,[13] described in terms of a general framework of birth, grow and decay of States or dynasties (*dawla*) determined by economic and social factors. The growth of culture, crafts and sciences depends on the demands made by luxury and wealth in all civilisations. According to Fernand Braudel's terminology Ibn Khaldūn's macro dimensions would lie between long "cyclical phases" and secular *longue durée* processes. However, Khaldūn's cycles are repetitive and therefore coincide with Braudel's *longue durée* and very *longue durée* periods.[14]

---

12 The expression "philosophy of history" is modern. As we will see farther in this article the term "philosophy" (*falsafa*) has a quite pejorative meaning in Ibn Khaldūn's exposition.

13 "It should be known that according to physicians and astrologers the natural life span of a person is one hundred and twenty years ... The same is the case with the life span of States ... However, usually States do not last beyond the life span of three generations", Ibn Khaldūn (1970), I, 305–306; cfr. Ibn Khaldūn (1980), I, 343.

14 Braudel (2009), first edition in French 1958.

فإذا تمدنت المدنية و تزيدت فيها الأعمال و وفت بالضروري و زادت عليه صرف الزائد حينئذٍ إلى الكمالات من المعاش ثم ان الصنائع و العلوم إنما هي للإنسان من حيث فكره الذي يتميز به عن الحيوانات و القوت له من حيث الحيوانية و الغذائية فهو متقدم لضرورته على العلوم و الصنائع و هي متأخرة عن الضروري و على مقدار عمران البلد تكون جودة الصنائع للتأنق فيها حينئذٍ و جودة ما يطلب منها بحسب دواعي الترف و الثروة[15]

> When a civilisation becomes refined, a surplus of labor appears in it and abundantly satisfies the necessities of the people. The surplus is then spent on luxurious activities.
>
> Arts and sciences are the product of man's intelligence, which distinguishes him from animals. His desire for nutrition derives from his animal and nutritive nature and comes before his needs for arts and sciences which, in their turn, come after [physical] necessities. The greatness of the civilisation of a country can be measured with the refinement of its arts and the quality of the demands these arts must satisfy according to the needs of its luxury and opulence.[16]

In this article we consider both levels of Ibn Khaldūn's exposition on history but we focus on the first, the axiomatic and technical one, which has drawn less scholarly attention to date. The law of *muṭābaqa* applies above all to this level. The historian works on sources and must verify whether it is inherently possible (*imkān*) or impossible (*istiḥāla*)[17] for the event to have occurred (*uqūʿ*). We will see farther in this article that the expression inherent "possibility" does not mean probability, an option among others. It rather means necessity, given that it fits into a finalistic conception of causality that derives from the Aristotelian theory of becoming and motion. An event happens because the principle (*aṣl*) inherent in its matter (*mādda*) predisposes it to becoming actual. Ibn Khaldūn combines Aristotelian physics with his materialistic conception of society. Reported events that are inherently possible shall be accepted by the historian, who will then verify their coincidence with external conditions. Conditions in their turn can attach themselves to the essence (*dhāt*) of civilisation (*ʿumrān*), can be an accidental (*ʿāriḍ*) to it or cannot possibly attach themselves to it. The

15 Ibn Khaldūn (1970), II, 307.

16 Ibn Khaldūn (1970), II, 307; cfr. Ibn Khaldūn (1980), II, 347.

17 The term *imtināʿ* also occurs, cfr. below.

distinction between essence (*dhāt*) and accident (*ʿaraḍ*), says Khaldūn, is the central issue of all sciences, the religious and the rational ones.

أما الأخبار عن الواقعات فلا بد في صدقها و صحتها من اعتبار المطابقة فلذلك وجب أن ننظر في إمكان وقوعه ... إذا كان ذلك فالقانون في تمييز الحق من الباطل في الأخبار بالإمكان و الاستحالة أن ننظر في الاجتماع البشري الذي هو العمران و نميز ما يلحقه من الأحوال لذاته و بمقتضى طبعه و ما يكون عارضاً لا يعتد به و ما لا يمكن أن يعرض له و إذا فعلنا ذلك كان لنا قانوناً في تمييز الحق من الباطل في الأخبار و الصدق و الكذب بوجه برهاني لا مدخل للشك فيه و حينئذ فإذا سمعنا عن شيء من الأحوال الواقعة في العمران علمنا ما نحكم بقبوله مما نحكم بتزييفه و كان لنا ذلك معياراً صحيحاً يتحرى به المؤرخون طريق الصدق و الصواب فيما ينقلونه و هذا هو غرض هذا الكتاب الأول من تأليفنا[18]

> As far as information about [historical] facts is concerned, it is necessary to consider their veracity and exactitude from the point of view of the correspondence [between them and their circumstances]. One must carefully consider whether they can possibly have happened … Therefore the norm for the distinction between reality and vanity in historical information consists in a careful study of human society (that is to say, civilisation) and in the distinction between [historical] circumstances that are inherent in the essence of the society, in conformity with its nature, other circumstances that are accidental and should not be taken into consideration, and [yet other circumstances] that cannot have existed at all.
>
> In so doing, we have an unfailing norm based on a demonstrative [method] for distinguishing reality from vanity and veracity from falsehood in historical information. Thus, when we hear about circumstances in which facts have taken place in the [history of] civilisation, we shall know which ones we can accept and which ones we must consider fictitious. We shall have a valid criterion by which historians can pursue the path of truth and exactitude in what they hand down to us. This is the very purpose of this first book.[19]

We would like to compare the passage quoted before in which Ibn Khaldūn refers to historical events in terms of their possibility and reality, with the well-

18 Ibn Khaldūn (1970), I, 61.
19 Cfr. Ibn Khaldūn, (1980), I, 76–77.

known passage of Aristotle's *Poetics* in which the Greek philosopher states that poetry is "more philosophical" (and therefore superior) than history because it deals with what is possible, while history deals with what has really occurred. Therefore, says Aristotle, poetry deals with the universal, history with the particular.

> φανερὸν δὲ τῶν εἰρημένων καὶ ὅτι οὐ τὸ τὰ γενόμενα λέγειν, τοῦτο ποιητοῦ ἔργον ἐστίν, ἀλλ' οἷα ἂν γένοιτο καὶ τὰ δυνατὰ κατὰ τὸ εἰκὸς ἢ τὸ ἀναγκαῖον. ὁ γὰρ ἱστορικὸς καὶ ὁ ποιητὴς οὐ τῷ ἢ ἔμμετρα λέγειν ἢ ἄμετρα διαφέρουσιν (εἴη γὰρ ἂν τὰ Ἡροδότου εἰς μέτρα τεθῆναι καὶ οὐδὲν ἧττον ἂν εἴη ἱστορία τις μετὰ μέτρου ἢ ἄνευ μέτρων): ἀλλὰ τούτῳ διαφέρει, τῷ τὸν μὲν τὰ γενόμενα λέγειν, τὸν δὲ οἷα ἂν γένοιτο. διὸ καὶ φιλοσοφώτερον καὶ σπουδαιότερον ποίησις ἱστορίας ἐστίν: ἡ μὲν γὰρ ποίησις μᾶλλον τὰ καθόλου, ἡ δ' ἱστορία τὰ καθ' ἕκαστον λέγει.[20]

> It is, moreover, evident from what has been said, that it is not the function of the poet to relate what has happened, but what may happen, what is possible according to the law of probability or necessity. The poet and the historian differ not by writing in verse or in prose. The work of Herodotus might be put into verse, and it would still be a species of history, with metre no less than without it. The true difference is that one relates what has happened, the other what may happen. Poetry, therefore, is a more philosophical and a higher thing than history: for poetry tends to express the universal, history the particular.[21]

Unlike Aristotle, Ibn Khaldūn's purpose in the science of history is to combine the particular and the universal, reality and truth. One of the reasons why untruth deforms historical information is that many historians ignore the nature of conditions (*ṭabāʾiʿ al-aḥwāl*), and what is substantial or accidental in them.

> و من الأسباب المقتضية له أيضاً و هي سابقة على جميع ما تقدم الجهل بطبائع الأحوال في العمران فإن كل حادث من الحوادث ذاتاً كان أو فعلاً لا بد له من طبيعة تخصه في ذاته و

20 Aristotle [c], *Poetics* 1451a37–1451b1–7 (1902), 34, where the Greek text is the same as Aristotle [b] (1965).

21 Aristotle [c], *Poetics* (1902), 35.

فيما يعرض له من أحواله فإذا كان السامع عارفاً بطبائع الحوادث و الأحوال ... أعانه ذلك في تمحيص الخبر على تمييز الصدق من الكذب[22]

> Another reason why [falsehood inevitably affects historical information], and this reason prevails over the other reasons mentioned, is ignorance of the nature of circumstances in [the history] of civilisation. Every event, be it an essence or a fact, necessarily has a specific nature inherent in its essence and in the circumstances that affect it. If the historian is aware of the nature of events and circumstances … this awareness will assist him in distinguishing truth from falsehood in the analysis of that [historical] information.[23]

The comparison with Aristotle's *Poetics* allows us to highlight better Ibn Khaldūn's purpose to get rid of philosophical abstraction and, thanks to a rigorous demonstrative method (*bi wajhin burhānin*),[24] to find out the general law of history, empirically verifiable: the law of *muṭābaqa*. Ibn Khaldūn quotes concrete examples of *muṭābaqa* at the *courte durée* and single event level. A concrete example of *mutābaqa* at the level of single events is the passage in which Ibn Khaldūn dismisses for lack of "coincidence" the anecdote quoted by Ibn ʿAbd Rabbih (860–940) in his *adab* treatise *Al-ʿIqd al-Farīd* ("The Unique Neckless")[25] about immoral and lascivious behavior of the caliph al-Ma'mūn (r. 813–833), which does not "conform" with social and psychological "conditions" of al-Ma'mūn himself.

و أين هذا كله من حال المأمون المعروفة في دينه و علمه و اقتفائه سنن الخلفاء الراشدين من آبائه و أخذه بسيرة الخلفاء الأربعة أركان الملة و مناظرته العلماء و حفظه لحدود الله في صلواته و أحكامه فكيف تصح عنه أحوال الفساق المستهترين في التطواف بالليل و طروق المنازل و غشيان السمر سبيل عشاق الأعراب[26]

22 Ibn Khaldūn (1970), I, 57–58.

23 Cfr. Ibn Khaldūn (1980), I, 72–73.

24 Ibn Khaldūn (1970), I, 61.

25 According to Pellat, *adab* prose treatises are of three categories: parenetic *adab* (ethical writings); cultural *adab* (prose or poetry fragments and anecdotes) suitable to be used in well-mannered society; training *adab*, consisted (handbooks for members of the ruling classes), Pellat, "Adab ii. Adab in Arabic Literature" (1983), 439–444.

26 Ibn Khaldūn (1970), I, 28–29.

> How does this story match the personality of al-Maʾmūn, his religion and learning, and the way he conformed to the behaviour of the righteous ʿAbbasid caliphs, his forefathers, and to the way of life of the first four caliphs, the pillars of religion? How does it match al-Maʾmūn's emulation of the religious scholars and his observance of the laws of God most High in his prayers and in his policies implementing the commandments of God? How could it be true that he behaved like a dissolute and shameless man who wanders around the city at night and enters other people's houses in the dark, in the manner of Arab Bedouin lovers?[27]

There is only one passage in which Ibn Khaldūn indirectly refers to the law of *muṭābaqa* at the *longue durée* level. It is a passage describing the general process of growth and decay of civilisations. Ibn Khaldūn observes that the historian cannot rely upon astrology in order to explain these phenomena. Astrologers, he says, explain social prosperity with a favorable "coincidence" between the stars and worldly affairs (*al-muṭābaqa bayna al-aḥkām al-nujūmiyya wa al-aḥwāl al-arḍiyya*).[28] The historian, instead, needs to find worldly causes (*al-sabab al-arḍī*) such as the density of the population and the available labour (*aʿmāl*), from which profit (*kasb*) results. Available surplus leads to prosperity of trade, crafts and sciences but also to an augmentation of taxes and State's revenues. This process in its turn produces inevitably an exponential augmentation of State's luxurious expenses and eventually provokes a chain reaction of decay: impoverishment of the population, fall of demand, crisis of underproduction and bankruptcy of the State. Then civilisation regresses to a nomad stage, or disappears.[29]

The *muṭābaqa* is an eloquent example of intersections between different disciplines and fields. These intersections have affected all aspects of human thought throughout history, including the development of modern scientific thought. Until the end of the seventeenth century, after Europe had engaged in its own appropriation of Arabic and Hellenistic mathematics and physics as early as the twelfth century, continuity, complexity and exchanges explain better the birth of modern scientific thought than scientific revolutions. Alexandre Koyré defined the modern world the "world of precision" as opposed to the ancient "world of approximation".[30] Historians of science agree that one of

---

27 Cfr. Ibn Khaldūn (1980), I, 39–40.

28 Ibn Khaldūn (1970), II, 246. Cfr. Ibn Khaldūn (1980), II, 282.

29 Economic cycles are more extensively described Ibn Khaldūn (1970), I, 108–113. Cfr. Ibn Khaldūn (1980), II, 118–124.

30 Koyré (1948).

the advances of the Arab-Islamic scientists was their highest precision in comparison with their Hellenistic predecessors, for example in the field of astronomical calculations.[31]

The notion of *muṭābaqa* in Ibn Khaldūn's historiography responds to his need for an instrument of the highest precision for the measurement of historical events and processes. However, Ibn Khaldūn's aspiration to precision could not but remain partially unsatisfied, because history is a social science and not a natural one. This is why the law of *muṭābaqa* cannot concretely play the role of an instrument of the highest precision for measuring events and conditions as it is presented in the programmatic pages of the *Muqaddima*. *Muṭābaqa* is a notion in the intersection between different fields and disciplines. It is a microcosm that contains much of the *Muqaddima* itself. The *Muqaddima*, as we have already pointed out, is an outstanding synthesis of the Arab classical cultural system from the point of view of Khaldūn's philosophy of history. For its being a "system", the *Muqaddima* contains aporias, that is to say contradictions that undermine the system itself. The fundamental aporia of Khaldūn's philosophy of history is to be found in the aprioristic and finalistic assumption of the law of *muṭābaqa*. Events happen in virtue of attributes inherent in the nature of things, which make their potential happening (*imkān*) necessary.

In a previous article of ours, we argued that Ibn Khaldūn's views on poetry are dominated by a Neo-Platonic paradigm, which he derived from Avicenna's *Psychology* (*'Ilm al-Nafs*).[32] Already Aristotle, in his *Psychology* (Gr. *Perì Psykhēs*, Lat. *De Anima*), despite his intention to get rid of Platonic ideas as essences separated from external reality, ends up saying that the universals "somehow" [pre]exist in the soul itself.

> αἴτιον δ' ὅτι τῶν καθ' ἕκαστον ἡ κατ' ἐνέργειαν αἴσθησις, ἡ δ' ἐπιστήμη τῶν καθόλου· ταῦτα δ' ἐν αὐτῇ πώς ἐστι τῇ ψυχῇ.[33]
>
> The reason is that what the sensations apprehends is the particular, while what knowledge apprehends is the universal, and the universal somehow [pre-]exists in the soul.[34]

Avicenna, an Aristotelian philosopher inclined to Neo-Platonism and very present in the *Muqaddima*,[35] deepened further this aspect of Aristotle's theory

31 Van Dalen (2007), 101–103.
32 Lelli (2014), 196–197.
33 Aristotle [d], *De Anima*, 417b12–13 (1902), 74. Same edition as Aristotle [e] (1961).
34 Cfr. Aristotle [d], *De Anima* (1902), 75.
35 Analogies between Avicenna's and Ibn Khaldūn's thought with regard to their respective theories of prophecy have been pointed out by Abdelali Elamrani (2009), 69–77.

of knowledge, particularly in his theory of prophecy, which he considered the highest degree of human capacity to know the universals through an intuitive and illuminative process of unity with the agent Intellect.

فمكن اذا ان يكون شخص من الناس مؤيد النفس لشدة الصفاء و شدة الاتصال بالمبادئ العقلية الى ان يشتعل حدسا اعني قبولا لها من العقل الفعال في كل شيء و ترتسم فيه الصورة التي في العقل الفعال اما دفعة و اما قريبا من دفعة[36]

> Il est possible qu'il y ait parmi les hommes un individu à l'âme fortifiée par la grande pureté et par l'étroite jonction avec les principes intellectuels jusqu'à ce qu'il s'enflamme d'une intuition intellectuelle, je veux dire en recevant les principes intellectuels de l'Intellect agent pour toute chose, et qu'en lui s'imprime la forme qui est dans l'Intellect agent, soit tout d'un coup, soit presque tout d'un coup …[37]

Given the materialistic character of Ibn Khaldūn's historiography, we infer that the term *muṭābaqa* does not refer to a transcendent union with the agent Intellect, but to a worldly knowledge at the level of sensual entities and the first intelligibles (*al-ma'qūlalāt al-uwwal*). Ibn Khaldūn turns a Neo-Platonic epistemological proceeding into an immanent law of history. He emancipates it from transcendence but not from finalism and apriorism.

In order to understand better the aporia of Ibn Khaldūn's historiography, we can compare it with what Umberto Eco called the "aporia" of the aesthetics of Thomas Aquinas (1225–1274). Eco inferred that according to Thomas Aquinas' theological views the only true beauty is the beauty of nature, and not the beauty of artistic products. However, only God and the angels can fully understand the beauty of the creation, while man can fully understand only the beauty of forms that have been produced by men, that is to say the beauty of artistic products. According to Eco this aporia undermined Thomas Aquinas' aesthetic system.[38]

Eco pursues the argument saying that the culture of Humanism in the "fifteenth century exploited that anomaly and disintegrated the aesthetic system of Thomas Aquinas from within. Neo-Platonism, which was rediscovered by

36 Avicenna [a], *Psychology* (1956), I, 246.
37 Avicenna [b], *Psychology* (1956), II, 177.
38 Eco (1993), 212.

Humanism, gave new dignity to the artist and his products, supported by the emergence of new urban classes proud of their initiative and social creativity".[39]

We think that a similar reasoning can be applied to the notion of *muṭābaqa* in Ibn Khaldūn's historiography, although he was far from living in an epoch of humanist renaissance. Neo-Platonism was the best resource at his disposal for giving coherence to his mechanistic and materialistic conception of history: an immanent kind of Neo-Platonism brought down to earth.

Ibn Khaldūn was perfectly aware of the novelty of his method. Indeed, after his definition quoted before of *muṭābaqa* as a "law" (*qānūn*) and "criterion" (*miʿyār*) for distinguishing right from wrong in historical reports, Ibn Khaldūn continues proudly claiming that he has established a "new science" (*mustaḥdath al-ṣanʿa, musṭanbaṭ al-nash'a*). He does not give a place to it in the classification of the sciences which, according to an Arab classical framework, he divides into philosophical (*ḥikmiyya falsafiyya*) and religious transmitted sciences (*naqliyya*). History is a branch of "wisdom" (*ḥikma*), where *ḥikma* apparently does not refer to the Arab-Islamic philosophy, but to a broader universal knowledge that is not related to a specific people, religion or time. History, he says, may seem similar to rhetoric (*ʿilm al-khiṭāba*) and politics (*ʿilm al-siyāsa al-madanīyya*), but it remains unclassifiable. It is an independent (*mustaqill*) and extraordinary (*gharīb al-nazʿa*) science, with its own peculiar subject and problems, extremely powerful (*ʿazīz*) and useful.[40]

Ibn Khaldūn's definition of history as a "new science" induced some scholars to compare the *Muqaddima* to the *Scienza Nuova* by Giambattista Vico (1668–1744).[41] We think that the most important analogy between Ibn Khaldūn and Vico consists in their aesthetic perception of history, but the method and purpose of their historiography is different. They have both an organicist conception of history, but while Ibn Khaldūn refutes legends and phantasies because they divert the historian form the objective and scientific truth, Vico highlights the historiographic value of subjective and irrational perceptions that ancient peoples expressed in their myths and poems.

Ibn Khaldūn's *muṭābaqa* is a rule that allows the historian to avoid untruth that "naturally affects historical information". There are several reasons why

39 Eco (1982), 2–3.

40 Ibn Khaldūn (1970), I, 61–62. Cfr. Ibn Khaldūn (1980), I, 77–78.

41 Gabrieli (1975), 122–126. There are several other of publications comparing these two authors. Elsewhere Gabrieli remarks that similarities between Ibn Khaldūn and Vico should not be exaggerated, Gabrieli (1967), 241, n. 1.

untruth affects historiography. One of them is the fact that people tend to trust the authority of transmitters (*nāqilūna*).[42] Criticism for blind trust in transmitters had been expressed with a similar programmatic strength by the mathematician and physicist Ibn al-Haytham (965–1040), author of the *Book of Optics* (*Kitāb al-manāẓir*),[43] which has been the reference in optics until the publication of Newton's *Principia* (1687). More than with Vico, Ibn Khaldūn might be compared with Ibn al-Haytham and not only because of their common warning against blind trust in transmitters.

Ibn al-Haytham developed a theory of light separated from his theory of vision. He described his theory of light, which was associated with geometrical and physical optics, according to a mathematical model based on experiments. His theory of vision, which was associated with the physiology of the eye and psychology of perception, rejected the theory of the visual ray emanating for the eye, and accepted the theory of the philosophers, like Avicenna. However, unlike the latter, he did not consider perceived forms as totalities but as the result of straight rays radiating toward the eye from every point of external objects, thanks to action of light.[44]

Ibn Khaldūn, despite his criticism of Hellenistic philosophy, accepts the main features of Aristotle's theory of vision, most probably relying upon Avicenna's psychology.[45] In the chapter on "The real meaning of prophecy" Ibn Khaldūn describes vision as a part of the process of knowledge that starts with the sensual perception, the faculty that "perceives sensual things with the organ of vision, hearing and all the other organs",[46] and transmits them to the other faculties hierarchically organised. Vision therefore consists of forms (*ṣuwar*) that the human soul draws from sensual objects, stores and combine them in the different organs that compose it: the senses, memory, the intellect.

Ibn Khaldūn applies a similar model to the process of poetical composition which he describes therefore as a particular case of cognition.[47] Despite the

42 Ibn Khaldūn (1970), I, 57. Cfr. Ibn Khaldūn (1980), I, 72.

43 Ibn al-Haytham [c], *Kitāb al-manāẓir* (1989). "The seeker after the truth is not one who studies the writings of the ancients ... puts his trust in them, but rather the one who suspects his faith in them and questions what he gathers from them, the one who submits to argument and demonstration, and not to the sayings of a human being whose nature is fraught with all kinds of imperfection and deficiency", Ibn al-Haytham (1989), 3, quoted by T.V. Venkateswaran (2016), 362.

44 Rashed, "Ibn al-Haytham: between mathematics and physics", in this volume.

45 Ibn Khaldūn (1970), I, 173 f. Cfr. Ibn Khaldūn (1980), I, 194 f.

46 Ibn Khaldūn (1970), I, 175. Cfr. Ibn Khaldūn (1980), I, 196.

47 Lelli (2014), 210. We note that Ibn Khaldūn calls coincidence between mental forms and sensual words combinations in poetical composition "*inṭibāq*", and not "*muṭābaqa*".

evident divergence between Ibn al-Haytham's theory of vision and the Aristotelian one, Ibn Khaldūn seems to assume that they can perfectly coexist. In fact, among the philosophical sciences, Ibn Khaldūn "refutes" physics and metaphysics but not mathematics, to which Ibn al-Haytham's optics formally belongs, being a branch of geometry. We have already insisted on the systematic character of the *Muqaddima*, a book that integrates all possible aspects of the Arab classical culture, to the maximum extent.

[المناظرة] و هو علم يتبين به أسباب الغلط في الإدراك البصري ... بناء على أن إدراك البصر يكون بمخروط شعاعي رأسه نقطة الباصر و قاعدته المرئي ثم يقع الغلط كثيرا في رؤية القريب كبيرا و البعيد صغيرا و كذا رؤية الأشباح الصغيرة تحت الماء و وراء الأجسام الشفافة كبيرة ... أمثال ذلك فيتبين في هذا العلم أسباب ذلك و كيفياته بالبراهين الهندسية ... و قد ألف في هذا الفن كثير من اليونانيين و أشهر من ألف فيه من الإسلاميين ابن الهيثم ... وهو من العلوم الرياضية و تفاريعها[48]

> [Another subdivision of geometry is optics.] It explains the reasons for errors in visual perception ... [which] takes place through a cone formed by rays, the top of which is the point of vision and the base of which is the object seen. Error often occur. Nearby things appear large. Things that are far away appear small ... small objects appear large under water or behind transparent bodies ... and so on. This discipline explains with geometrical proofs the reasons for these things ... Many Greek scholars have written on this subject, while the most famous Muslim author on optics is Ibn al-Haytham ... It is a branch of the mathematical sciences.[49]

Despite a certain degree of incoherence due to Ibn Khaldūn's *inclusive* attitude, there is indeed a common element between Ibn al-Haytham's optics and Ibn Khaldūn's historiography, more important than their analogous claim for emancipation from the authority of transmitters. Ibn al-Haytham studied the causes (*asbāb*) and the how (*kayfiyyāt*) of vision through geometrical demonstration (*bi al-barāhīn al-handasiyya*) thus overcoming the errors of perception (*idrāk*). Ibn Khaldūn studies the how (*kayfiyyāt*)) and why (*asbāb*) of events

48 Ibn Khaldūn (1970), III, 104.
49 Ibn Khaldūn (1980), III, 132–133.

through a demonstrative method (*bi wajhin burhānin*) thus overcoming the appearance (*ẓāhir*) of history as mere chronicle of events.

> إذ هو في ظاهره لا يزيد على أخبار عن الأيام والدول والسوابق من القرون الأول تنمق لها الأقوال و تضرب فيها الأمثال و تطرف بها الأندية إذ غصها الاحتفال ... وفي باطنه نظر وتحقيق وتعليل للكائنات ومبادئها دقيق وعلم بكيفيات الوقائع وأسبابها عميق فهو لذلك أصل في الحكمة عريق[50]

> In appearance, history is nothing more than a report on past events and dynasties, ancient facts coloured with anecdotes and proverbs with the purpose of entertaining gatherings ... However, the inner nature of history is its being an attentive study, an accurate investigation and an explanation of existing things and their origins, and a profound knowledge of the characters and causes of events. Therefore history is an ancient branch of wisdom[51]

In conclusion of this section, the law of *muṭābaqa* designates at once an objective law of history and the scientific method of the historian, explicitly applicable to *courte durée* phenomena, namely single events. We have seen that despite Khaldūn's materialistic and empiric conception of history, the law of *muṭābaqa* is affected by a finalistic and aprioristic aporia. The historian practices empiric observation of reality and sources, to discover that historical events are predetermined by attributes inherent in their substance. Once they become actual, events necessarily "coincide" with conditions.

2.2 Muṭābaqa *and Occult Sciences. The* zā'īraja, *a Divination Technique*

In this section we analyse the occurrence of *muṭābaqa* in passages of the *Muqaddima* devoted to occult sciences which wrongly pretend to prove the truthfulness of their deductions by referring to a kind of "coincidence" that is in fact fictitious and purely tautological.

In the section of the *Muqaddima* devoted to "the various kinds of human beings who have supernatural perception" ("*fī aṣnāf al-mudrikīna li-al-ghayb*"),[52] Ibn Khaldūn distinguishes between truthful and false perceptions.

---

50 Cfr. Ibn Khaldūn (1980), I, 6.

51 Cfr. Ibn Khaldūn (1980), I, 6.

52 Ibn Khaldūn (1970), I, 165f. Ibn Khaldūn (1980), I, 184f.

Prophecy, holiness and certain dream visions are truthful perceptions of the supernatural. The highest supernatural perception is prophecy. The Qur'ān is the greatest miracle (*muʿjiza*) and its truthfulness does not need to be proved by the coincidence of its words with external reality, because there is a perfect identity (*ittiḥād*) in it between the proof (*dalīl*) and what needs to be proved (*madlūl*). The word *ittiḥād* (litt. "unity") in this context is synonymous of *muṭābaqa*.

> أعظم المعجزات و أشرفها و أوضحها دلالة القرآن الكريم المنزل على نبينا صلوات الله و سلامه عليه لأن الخوارق في الغالب تقع مغايرةً للوحي الذي يتلقاه النبي و تأتي المعجزة شاهدة به و هذا ظاهرو القرآن هو بنفسه الوحي المدعا و هو الخارق المعجز و دلالته في عينه و لا يفتقر إلى دليل أجنبي عنه كسائر الخوارق مع الوحي فهو أوضح دلالة لاتحاد الدليل و المدلول فيه[53]
>
> The greatest, noblest and clearest miracle is the evidence which is the Qur'ān itself, revealed to our Prophet, God bless him and give him salvation! Indeed, usually wonders [*khawāriq*] are separated from the prophetic revelation, and miracles take place to testify to the truthfulness of that revelation. This is evident.
>
> The Qur'ān, instead, is at once a revelation and a miraculous wonder. It is the evidence of itself and does not need any external proof, as is the case for other wonders which take place in concomitancy with revelation. The Qur'ān is the clearest evidence for it is both the proof and the object to be proved.[54]

Ibn Khaldūn mentions explicitly *muṭābaqa* as perfect identity between subject and object of knowledge in the case of the angels. They are incorporeal essences, pure intelligences, in which there is a perfect unity between intellect, thinker and object of thinking (*al-ʿaql wa al-ʿāqil wa al-mʿaqūl*). There is a perfect "coincidence" (*muṭābaqa*) between their knowledge (*ʿulūmihim*) and the things to be known (*maʿlūmātihim*).

> للائكة الذين ذواتهم من جنس ذواته و هي ذوات مجردة عن الجسمانية و المادة و عقل صرف يتحد فيه العقل و العاقل و المعقول و كأنه ذات حقيقتها الإدراك و العقل فعلومهم

53 Ibn Khaldūn (1970), I, 171.

54 Cfr. Ibn Khaldūn (1980), I, 191–192.

حاصلة دائما مطابقة بالطبع لمعلوماتهم لا يقع فيها خلل البتة و علم البشر هو حصول صورة المعلوم في ذواتهم بعد أن لا تكون حاصلة فهو كله مكتسب[55]

> The essences of the angels are of the same kind of the essences [of the intellectual and the spiritual world, *'ālam al-'aql wa al-arwāḥ*] ... They are essences abstracted from corporeality and matter, they are pure intellect in which the intellect, the one who knows and the object of knowing become one. You would say they are essences the reality of which is perception and intellect [at one and the same time]. Their knowledge is always characterised by a natural coincidence [*muṭābaqa*] with the object of their knowing, without any imperfection. The knowledge of the human beings, instead, consists in attaining the form of the object of knowing in their essences, after it had not been there. It is a totally acquired type of knowledge.[56]

Most of divination techniques use tricks, therefore when there seem to be a conformity between what needs to be proved and external reality, it is either false or the result of a fortuitous "coincidence". The most interesting example of *muṭābaqa* in occult sciences has to be found in a divination technique called "*zā'īraja* of the world", the invention of which is attributed to Abū al-'Abbās al-Sabtī, a Maghrebi mystic who according to Ibn Khaldūn lived in the twelfth century. This technique uses a table in which are represented the celestial spheres, alphabetical letters, numbers and obscure riddles. The person who consults the soothsayer asks a question about the future. The soothsayer combines different elements represented in the table and gives an answer which he pretends to be a correspondence (*muṭābaqa*) between the question and a future event.

Ibn Khaldūn denies all credibility to this technique because supernatural perceptions are a natural gift of God given to prophets. They can also be acquired by mystics (*ahl al-ryāḍa*) through mystical exercises. Both prophets and mystics can go beyond reason. Supernatural perceptions cannot be acquired as a trivial technique (*ṣinā'a*).

Ibn Khaldūn puts forward other arguments which highlight further that the notion of *muṭābaqa* lies in the intersection between the classical Arab system of thought and a new one. He says that in worldly sciences agreement between entities (*al-tanāsub bayna al-umūr*) is truthful only if it provides knowledge of

---

55 Ibn Khaldūn (1970), II, 371–372.
56 Cfr. Ibn Khaldūn (1980), II, 421.

an actual event that exists either in reality or in science (*al-wāqiʿāt al-ḥāṣila fī al-wujūd aw fī al-ʿilm*). It does not provide knowledge of future entities, unless their causes (*asbāb wuqūʿihā*) are known.

Apparently Ibn Khaldūn refers here to the logic proceeding of *qiyās* (analogy), which consists of deducing the known from the unknown (*majhūluhā min maʿlūmihā*). Divination techniques like the *zāʾīraja* can realise only tautological agreements between the question asked by the person who consults the soothsayer and the answer given by the latter as a result of manipulation of letters or numbers. This answer has nothing to do with necessary coincidence between speech and external reality (*muṭābaqat al-kalām li mā fi al-khārij*).

فظهر أن التناسب بين الأمور هو الذي يخرج مجهولها من معلومها و هذا إنما هو في الواقعات الحاصلة في الوجود أو العلم و أما الكائنات المستقبلة إذا لم نعلمْ أسباب وقوعها و لا تثبت لها خبر صادق عنه فهو غيب لا يمكن معرفته و إذا تبين لك ذلك فالأعمال الواقعة في هذه الزايرجة كلها إنما هي استخراج ألفاظ الجواب من ألفاظ السؤال لأنها كما رأيته استنباط حروف على ترتيب من تلك الحروف بعينها على ترتيب آخر و سر ذلك إنما هو من تناسب بينهما يطلع عليه بعض دون بعض فمن عرف ذلك التناسب تيسّر عليه استخراج ذلك الجواب بتلك القوانين و الجواب يدل في مقام آخر من حيث وضوع ألفاظه و تراكيبه على وقوع أحد طرفي السؤال مِن نفي أو إثبات و ليس هذا من المقام الأول بل إنما يرجع الى مطابقة الكلام لما في الخارج و لا سبيل إلى معرفة ذلك من هذه الأعمال بل البشر محجوبون عنه و قد استأثر الله بعلمه و الله يعلم و أنتم لا تعلمون[57]

It is evident that the relationship of conformity between things allows one to deduce the unknown from the known. But this is true only for events occurring in the world of existence or in science, while things that will exist in the future can only be known if their causes are known and if we are truthfully informed about them, because [otherwise] they belong to the Unseen which cannot be known. Therefore it is clear that all the operations of the *zāʾīraja* consist in deducing the words of the answer from the words of the question. As you have seen, [this operation] consists in deducing a new sequence of letters from a given sequence of letters. The secret lies in a relationship of conformity between the two sequences. Some may

57 Ibn Khaldūn (1970), I, 219–220.

> have knowledge of it, someone else may not. Those who have knowledge of that relationship of conformity can easily deduce the answer by applying the fore-mentioned rules. In other cases the answer to the question can be a positive or negative statement with certain words' sequences and combinations. This is another kind of operation. [It does not simply consist in deducing the words of the answer from the question]. It implies coincidence between speech and external reality. But through the operations [of the *zā'īraja*] it is impossible to acquire such knowledge, which is hidden to the human beings. God claim His knowledge. "God knows and you don't know"[58]

Ibn Khaldūn gives a concrete example in which the soothsayer manipulates numbers and presents as agreement (*tanāsub*) between question and answer what is in fact the solution of an equation and has nothing to do with the supernatural and the knowledge of the future.

> و إذا كان كثير من المعاناة في العدد الذي هو أوضح الواضحات يعسر على الفهم إدراكه لبعد النسبة فيه و خفايها فما ظنك مثل هذا مع خفاء النسبة فيه و غرابتها (فلنذكر) مسئلة من المعاناة يتضح لك بها شيء مما ذكرنا مثاله لو قيل لك خذ عدداً من الدراهم و اجعل بإزاء كل درهم ثلاثة من الفلوس ثم أجمع الفلوس التي أخذت و اشتر بها طايراً ثم اشتر بالدراهم طيوراً بسعر ذلك الطاير فكم الطيور المشتراة فجوابه أن تقول هي تسعة لأنك تعلم أن فلوس الدراهم أربعة و عشرون و أن الثلاثة ثمنها و أن عدة أثمان الواحد ثمانية فكأنك جمعت الثمن من كل درهم إلى الثمن الآخر فكان كله ثمن طاير فهي ثمانية طيور عدة أثمان الواحد و تزيد على الثمانية طايراً آخر و هو المشترى بالفلوس المأخوذة أولاً و على سعره اشتريت بالدراهم فتكون تسعة فأنت ترى كيف خرج لك الجواب المضمر بسر التناسب الذي بين أعداد المسئلة و الوهم أول ما يلقي إليك هذه و أمثالها إنما يجعله من قبيل الغيب الذي لا يمكن معرفته[59]

> Many an operation with numbers, which are the clearest things in the world, is difficult to grasp, because the (existing) relations are difficult to establish and intricate. This is the case to a much greater degree here, where the relations are so intricate and strange. Let us mention a problem that will to some degree illustrate the point just stated.

58 Cfr. Ibn Khaldūn (1980), I, 245.

59 Ibn Khaldūn (1970), I, 218–219.

> Take a number of dirhams and place beside each dirham three *fals*. Then, take all the *fals* and buy a fowl with them. Then, buy fowls with all the dirhams for the same price that the first bird cost. How many fowls will you have bought?
>
> The answer is nine. As you know, a dirham has twenty-four *fals*, three *fals* are one-eighth of a dirham, one is eight times one-eighth. Adding up one-eighth of each dirham buys one fowl. This means eight fowls (for the dirhams), as one is eight times one-eighth. Add another fowl, the one that was bought originally for the additional fats and that determined the price of the fowls bought with the dirhams. This makes nine. It is clear how the unknown answer was implied in the relations that existed between the numerical data indicated in the problem. This and similar (things) are at first suspected as belonging to the realm of the supernatural, which cannot be known.[60]

Franz Rosenthal has translated this operation into modern mathematical symbols:

$$y\tfrac{1}{8} = 1$$
$$y + y\tfrac{1}{8} = x$$
$$x = 8 + 1 = 9$$

This example is interesting not only because it proves the "rational spirit" of Ibn Khaldūn,[61] but also because it means that the truth of numbers is not necessarily the truth of things, a fact that was well known to Arab-Islamic mathematicians. Al-Zanjānī (d. 1261 ca.) for instance in his treatise on algebra *Qisṭās al-muʿādala fī ʿilm al-jabr wa al-muqābala* states that numbers unlike the other words do not designate a reality (*ḥaqīqa*) but quantities (*miqdār*).

إذا أطلقوا بالعدد فلا يريدون به حقيقته و إنما يريدون مقداراً ما سواء كان واحداً أو أقل[62]

> Lorsqu'on désigne un nombre on n'entend pas sa vérité réelle mais plutôt une grandeur qui soit égale à une unité, ou moins ou plus.[63]

Although the ontological status of numbers would be discussed in depth only in the nineteenth and twentieth centuries, it is meaningful that Ibn Khal-

60 Ibn Khaldūn (1980), I, 244–245.
61 Ifrah (1994), 291.
62 Sammarchi (2017), I, 265.
63 Sammarchi (2017),) II, 513.

dūn uses this argument in order to highlight better that only historiography provides a scientific method that verifies hypotheses empirically, in external reality, when the law of *muṭābaqa* (between events and conditions) is correctly applied. Without empiric observation of reality, associated with a rigorous demonstrative reasoning, truth cannot be achieved. Worldly truth has neither a transcendent meaning nor an abstract logical one. It is the truth of sensual and historical reality, which exists in time (*zaman*, *ayyām*) and is subject to continuous change. Ibn Khaldūn insists that in order to avoid applying analogical reasoning in a tautological manner, the historian must take into the greatest consideration change in conditions (*tabaddul al-aḥwāl*) in time.[64]

Analogy is a valid instrument for understanding events only if it has at once a demonstrative and an empiric dimension, and not, as in the case of the equation quoted before, if it is tautological and separated from reality.

> و القياس و المحاكاة للإنسان طبيعة معروفة و من الغلط غير مأمونة تُخرِجه مع الذهول و الغَلط عن قصده و تعوج به عن مرامه فربما يسمع السامع كثيراً من أخبار الماضين و لا يتفطّن لما وقع من تغير الأحوال و انقلابها فيجريها لأول وَهْلة على ما عرف ويَقيسها بما شَهِد و قد يكون الفرق بينهما كثيراً فيقع في مهواة من الغلط[65]

> It is well-known that analogical reasoning and comparison are part of human nature. They are not exempt from mistake, together with absent-mindedness and error. They take man away from his purpose and sway him from his aim. Sometimes man receives a lot of historical information from the past but does not understand how conditions have changed profoundly. He immediately applies what he knows about the present to the history [of the past] and measures the latter by the events he has witnessed, although the difference between the [present and the past] may be great. Therefore he falls into an abyss of error.[66]

We can conclude this section saying that according to Ibn Khaldūn *muṭābaqa* does not apply to the knowledge of the supernatural or of future events the causes of which are unknown. This knowledge is indeed impossible for ordinary people. Analogy, syllogism and equations, if abstracted from reality are

64 Ibn Khaldūn (1970), I, 44. Cfr. Ibn Khaldūn (1980), I, 56.
65 Ibn Khaldūn (1970), I, 46.
66 Cfr. Ibn Khaldūn (1980), I, 58.

operations that can be formally correct but not true and it would be erroneous to associate reality with them.

2.3 Muṭābaqa *and Philosophy*

In this section we analyse the notion of *muṭāaqa* in the chapter of the *Muqaddima* on "The refutation of philosophy" (*fī ibṭāl al-falsafa*). We argue that despite his critical position vis-à-vis Islamic Hellenistic philosophy, Ibn Khaldūn ascribes to the historiographic law of *muṭābaqa* a philosophical content which does not solve the aprioristic and finalistic aporia that affects it.

The notion of *muṭāaqa* is a philosophical term and contributes answering epistemological questions: what is knowledge and what are its criteria and objects with regard to the worldly and social field? What is knowledge and what are its criteria and objects with regard to the theological and spiritual fields? As far as the latter are concerned, Ibn Khaldūn directs the seeker to theology and Sufism. As far as the worldly and social fields are concerned, we can infer that, according to Ibn Khaldūn, historiography is *the* science of the sciences, the science by excellence. The problem of worldly knowledge is that man, unlike the angels, faces a duality between two terms: himself, the knower, and the things to be known in external reality (*al-wujūd*). This duality can be solved through "coincidence" between these two terms.[67]

Ibn Khaldūn classifies the human beings in three categories, according to their cognitive capacities and sciences: scholars (*ʿulamāʾ*), the highest example of whom are probably the historians, men of mystical learning (*awliyāʾ*) and prophets (*anbiyāʾ*). "Scholars" do not have access to spiritual knowledge. Their capacities are limited to the knowledge of the primary *intelligibilia* (*al-maʿqūlāt al-uwwal*), to perceptive (*taṣawwur*) and apperceptive knowledge (*taṣdīq*).

النفوس البشرية في ذلك على ثلاثة أصناف صنف عاجز بالطبع من الوصول إلى الإدراك الروحاني فيقنع بالحركة إلى الجهة السُفلى نحو المدارك الحِسّية و الخيالية و تركيب المعاني من المحافظة الوهمية على قوانين محصورة و ترتيب خاص يستفيدون به العلوم التصورية و التصديقية التي للفكر في البدن و كلها خيالي منحصر نطاقه إذ هو من جهة مبدئه ينتهي إلى الأوليات و لا يتجاوزها و إن فسدت فسد ما بعدها و هذا هو في الأغلب نطاق الإدراك البشري الجسماني و إليه تنتهي مدارك العلماء و فيه تَرسُخ أقدامهم[68]

67 ʿĀbid al-Jabrī believes that the central issue of the whole *Muqaddima* is epistemology, cfr. ʿĀbid al-Jabrī, (2006), 323f.

68 Ibn Khaldūn (1970), I, 176–177.

> The human souls are of three kinds. The first is naturally incapable of attaining spiritual perception and he is satisfied to move in the lowest direction toward the sensual and imaginative perceptions and to compose the ideas stocked in the estimative memory according to given rules and specific arrangements. In this way people acquire perceptive and apperceptive knowledge which belongs to the intelligence and lies in the body. [This knowledge] is imaginative and limited in extent because from the way it begins it ends in the first intelligibles but cannot go beyond them. If it corrupts, everything beyond it will also be corrupted. The extent of human perception is corporeal. It is the end of the perceptions of scholars. Scholars are deeply rooted in it.[69]

Apparently the notion of *muṭābaqa* (and variants *inṭibāq, taṭbīq*) was well-rooted in Arab classical culture, given that it occurs in philosophy, linguistics and geometry. Ibn Khaldūn's use of it in historiography seems to be an innovation of his own. In the section of the *Muqaddima* devoted to Islamic philosophy, Ibn Khaldūn refutes the philosophers' claim to explain the physical and the transcendent world with the rational demonstrative method. He proves to be well acquainted with Aristotelian physics and metaphysics, most probably in the version of Avicenna, whose *Kitāb al-Shifā'* he quotes. He says indeed that physics (*al-ṭabīʿiyyāt*) studies bodies from the point of view of "the motion and stationariness. It studies celestial and elementary bodies and the human beings, the animals, the vegetables and the minerals that are generated from them".[70]

In other passages of the *Muqaddima* Ibn Khaldūn refers to becoming and change in terms of matter (*māddā*) acquiring a form (*ṣūra*). He uses the vocabulary of Aristotelian physics as the appropriate scientific vocabulary for describing historical processes. But does it really mean that he applies Aristotelian physics to history despite his refutation of Hellenistic philosophy? We think that the use of the same vocabulary does not mean real analogy in methods and content.

The reasons why Ibn Khaldūn refutes "philosophy" is that its results are not verifiable with the law of *muṭābaqa*. Philosophers, says Ibn Khaldūn, aim at understanding rationally the sensual and the transcendent world all together. They pretend that they can explain the inherent causal relations of these two worlds through logic.[71] However, he pursues, their demonstrations (*barāhīn*)

69 Cfr. Ibn Khaldūn (1980), I, 197–198.

70 Ibn Khaldūn (1970), III, 116. Cfr. Ibn Khaldūn (1980), III, 147.

71 Ibn Khaldūn (1970), III, 209–210. Cfr. Ibn Khaldūn (1980), III, 246–247.

are defective (*qāṣira*), and do not suffice to reach their goal. This is proved by the fact that coincidence (*muṭābaqa*) between their deductions (*natā'ij dhihniyya*) and external reality (*mā fī al-khārij*) cannot be verified. Their deductions instead are general (*ʿāmma*) and abstract, while external reality is material and individual (*mushakhkhaṣa bi mawāddihā*). In the sensual world coincidence between deductions and external reality must be verified through empiric observation (*shuhûd*). It cannot rely upon abstract logic.

Ibn Khaldūn mitigates his position by admitting that logic can grasp the first intelligibles (*al-maʿqūlāt al-uwwal*) by induction[72] and deduce simple truths from the individuals. However, he says, it is not recommended to attend to physics, a discipline that has nothing to do either with religion (*dīn*) or with economic life (*al-maʿāsh*, "making a living").[73] "*Al-maʿāsh*" is one of the main subjects of the science of history.

و[الفلاسفة] كأنهم في اقتصارهم على إثبات العقل فقط و الغفلة عما وراءه بمثابة الطبيعيين المقتصرين على إثبات الأجسام خاصة المعرضين عن النفس و العقل المعتقدين أنه ليس وراء الجسم في حكمة الوجود شيء و أما البراهين التي يزعمونها على مدعياتهم في الموجودات و يعرضونها على معيار المنطق و قانونه فهي قاصرة و غير وافية فيه بالغرض أما ما كان منها في الموجودات الجسمانية ويسمونه العلم الطبيعي فوجه قصوره أن المطابقة بين تلك النتائج الذهنية التي تستخرج بالحدود و الأقيسة كما في زعمهم و بين ما في الخارج غيريقيني لأن تلك أحكام ذهنية كلها عامة و الموجودات الخارجية متشخصة بموادها و لعل في المواد ما يمنع من مطابقة الذهني الكلي للخارجي الشخصي اللهم إلا ما يشهد له الحس من ذلك فدليله شهوده لا تلك البراهين فأين اليقين الذي يجدونه فيها و ربما يكون تصرف الذهن أيضا في المعقولات الأول المطابقة للشخصيات ... فيكون الحكم حينئذ يقينيا بمثابة المحسوسات إذ المعقولات الأول أقرب إلى مطابقة الخارج لكمال الانطباق فيها فتسلِّم لهم حينئذ دعاويهم في ذلك إلا أنه ينبغي

---

72 This is Avicenna's definition of first intelligibles: "J'entends par intelligibles premiers les prémisses auxquelles l'assentiment n'est pas donné par acquisition ni tandis que celui qui y adhère sent qu'il peut s'abstenir d'y adhérer un seul instant, comme notre croyance que le tout est le plus grand que la partie". "أعني بالمعقولات المقدمات التي يقع بها التصديق لا باكتساب و لا بأن يشعر المصدّق بها أنها يجوز له أن يخلو عن التصديق باه وقتاً ألبتة مثل اعتقادنا بأن الكل أعظم من الجزء", Goichon (1938), 235, quotes Avicenna [c], *Healing* [*Shifā'*] (1886), I, 292, and refers to Avicenna [d], *Deliverance* [*Najāt*] (1913), 270.

73 Ibn Khaldūn (1970), III, 214f. Ibn Khaldūn (1980), II, 309f.

لنا الإعراض النظر فيها إذ هو من ترك المسلم لما لا يعنيه فإن مسائل الطبيعيات لا تُهِمّنا في دينِنا ولا معاشنا فوجب علينا تركها[74]

The philosophers limit themselves to affirming the intellect and neglect what goes beyond it, in the same way as the physicists who limit themselves to affirming the body and disregard the soul and the intellect, believing that the wisdom inherent in the world of existence does not contemplate anything beyond the body.

The proofs [the philosophers] bring for their allegations about existing things saying that they rely upon the criterion of logical rules, are defectuous and do not reach their goal. The proofs they bring for their allegations about corporeal things constitute what they claim to be the subject of physics. However, the deficiency lies in the fact that coincidence [*muṭābaqa*] between the mental conclusions they pretend to draw from rational norms and analogical reasoning and external reality, are uncertain. The reason for that is that all mental judgments are general, while the external existing things are particular, as far as their material substance is concerned. There might be something in matter that prevents coincidence between universal mental judgments and particular external reality. This is not the case for sensual perception which is proved by empirical observation and not by logical arguments. Where is certainty in their proof, then?

Often the mind uses primary *intelligibilia* [*al-maʿqūlāt al-uwwal*] which conform [*muṭābaqa*] with the particulars… In this case the judgment is certain as in the case of judgments about sensual perception. The first *intelligibilia* are indeed more likely to coincide [*muṭābaqa*] [with external reality] because they conform [*inṭibāq*] perfectly with existing things.

Therefore we can accept the allegations [of the philosophers] in this respect, although we should refuse to attend to these matters because it is the duty of the Muslim to leave aside what does not concern him. In our religious and economic affairs we are not interested in physical questions.[75]

Philosophers, according to Ibn Khaldūn, do not only fail in the field of physics but also in the field of metaphysics (*mā baʿd al-ṭabīʿa*), because they pretend to deduce spiritual essences from the observation of external sensual real-

74 Ibn Khaldūn (1970), III, 213–214.

75 Cfr. Ibn Khaldūn (1980), III, 250–252.

ity, while the only intelligibles that are accessible to the senses (*fī mā huwwa madārik li nā*) can be deduced from experience of external reality. All that is immaterial cannot be proved by logic.

> و أما ما كان منها في الموجودات التي وراء الحس و هي الروحانيات ويسمونه العلم الإلهي و علم ما بعد الطبيعة فإن ذواتها مجهولة رأسا و لا يمكن التوصل إليها و لا البرهان عليها لأن تجريد المعقولات من الموجودات الخارجية الشخصية إنما هو ممكن فيما هو مدرك لنا بالحس فتنتزع منه الكليات و نحن لا ندرك الذوات الروحانية حتى نجرد منها ماهيات أخرى لحجاب الحس بيننا و بينها فلا يتأتّى لنا برهان عليها و لا مدرك لنا في إثبات وجودها على الجملة إلا ما نجده بين جنبينا من أمر النفس الإنسانية و أحوال مداركها و خصوصا في الرويا التي هي وجدانية لكل أحد و ما وراء ذلك من حقيقتها و صفاتها فأمر غامض لا سبيل إلى الوقوف عليه[76]

> The spiritual matters that go beyond the senses are the subject of the "divine science" or "metaphysics". Its essences are absolutely unknown, they cannot be grasped nor can they be proved logically. Indeed, an abstraction of the particular intelligibles that belong to the external world is possible only through the senses. The latter deduce the universals [from those particular intelligibles].
>
> We cannot perceive the spiritual essences until we abstract further quiddities from them because the senses are like a veil between us and them. We cannot find any logical proof about them nor can we have any cognition for affirming their existence.
>
> Within our inner being we can find the perceptions of the human soul and especially the dream visions which are intuitive for all. However, realities and qualities that go beyond [these perceptions] are obscure and there is no way to understand them.[77]

As far as we have seen, Ibn Khaldūn's refutation of Hellenistic philosophy perfectly fits in his views on historiography. It highlights further what he intends by *muṭābaqa* both as an objective historical law as a scientific method that guarantees certitude of knowledge.

Muṭābaqa in historiography appears to be a specific case of "coincidence" between the first intelligibles (the deductions of the historian) and material

76 Ibn Khaldūn (1970), III, 214–215.
77 Cfr. Ibn Khaldūn (1980), III, 252.

and individual external reality (historical events and processes). In order to ensure the scientific value of *muṭābaqa*, the historian must get rid of not only occult sciences and irrational practices associated with them, but also abstract rationalist philosophy. The law of *muṭābaqa* allows the historian to overcome dualism inherent in the act of knowing, a dualism that in the spiritual world of the angels does not subsist thanks to a perfect coincidence between subject and object, between their knowledge (*ʿulūmihim*) and the things to be known (*maʿlūmātihim*).

In conclusion of this section, we can say that despite his refutation of philosophy, Ibn Khaldūn gives a philosophical content to the historiographic law of *muṭāaqa*, a law that lies in the intersection between its being empirically observable and its being an axiom, that is to say a primary truth that cannot be demonstrated and from which demonstration derives. This is the meaning of the aprioristic aporia that affects it. Ibn Khaldūn aspires to a degree of precision which a social science like history cannot reach in reality. Only modern physics would be able to describe the laws of nature with the highest precision through mathematical laws.

2.4 Muṭābaqa *between Arab Linguistic Sciences and Logic*

In this section we analyse Khaldūn's use of *muṭābaqa* in parts of the *Muqaddima* dealing with the Arab linguistic sciences (*ʿulūm al-lisān al-ʿarabī*) and the impact of logic (*ʿilm al-manṭiq*) on it. In conformity with the Arab-Islamic tradition, Ibn Khaldūn considers the linguistic sciences being "auxiliary" of the religious sciences (*al-ʿulūm allatī tuhayyiʾuhā li al-istifāda*),[78] and logic being "instrumental" (*āliyya*) to the philosophical sciences.[79]

The above mentioned technical meaning of *muṭābaqa* which occurs in the *Kitāb al-ʿAyn* as "coincidence" between the two stones of a mill, suits very well the linguistic requirement of perfect "coincidence" between words (*lafẓ*) and meanings (*maʿānī*), between speech (*kalām*) and situations (*aḥwāl*). The search for precision in Arabic linguistic sciences is confirmed by the existence of fruitful exchanges between linguistics and mathematics, as in the case of the impact of linguistics on algebra and on the beginning of the combinatorial analysis highlighted by Roshdi Rashed in his analysis of the *Kitāb al-ʿAyn*.[80] Jamal Eddine Bencheikh in his book on Arab poetics remarked that for the literary scholar Ibn Quṭayba (828–889) there is *only one* uttering that perfectly trans-

78 Ibn Khaldūn (1970), II, 385. Cfr. Ibn Khaldūn (1980), II, 437.

79 On the classification of the sciences cfr. Jolivet (1996), 1008–1025.

80 Rashed (2011), 111–133.

lates into words a certain "meaning" (*maʿnā*), with the precision of an arrow that hits the target.

> Disons tout de suite que le *maʿnā* désigne la signification visée. La critique du sens organise dès lors une confrontation entre l'énoncé effectivement réalisé, l'énoncé idéal censé signifier parfaitement l'objet, et enfin l'objet lui-même que le poète a entrepris d'exprimer. L'énoncé idéal constitue une forme-sense abstraite, parfait et donc unique qui épuise d'un coup la signification visée. Si le poète arrive à la traduire en mots, il aura réalisé une opération mimétique: son énoncé se sert substitué à l'objet, sera devenu lui en quelque sorte … il a réalisé l'énoncé idéal, il a atteint la cible en plein centre.[81]

The requirement of conformity between speech and situation (*muqtaḍā al-ḥāl*) is a rule that dominated Arab eloquence until modern times. Ṭaha Ḥusayn admitted the use of colloquial Arabic in modern literature when "the situation required it", referring to al-Jāḥiẓ (776–869).[82] Ibn Khaldūn exposes the same requirement in passages of the *Muqaddima* on eloquence (*ʿilm al-balāgha*) and the linguistic sciences: the requirement of conformity (*muṭābaqa*) between speech and purpose, between words and meanings.

We have seen before that the *Kitāb al-ʿAyn* also mentions the meaning of *ṭabaqa* as "social group". Later, this meaning would also acquire a sociolinguistic sense. There are two main classes of people: the élite (*al-khāṣṣa*) and the masses (*al-ʿāmma*). The opposition is more cultural than socio-economic and refers to intellectual and moral qualities.

A similar notion of conformity between content, style and purpose of speech, with a similar sociolinguist background, can be found in Latin Europe where the *artes dictaminis* (techniques of prose composition) and *artes poeticae* (poetics) in the eleventh to thirteenth centuries relied upon the ancient notion of "convenient" (Greek "*prépon*", Latin "*aptum*"), meaning that speech needed to be convenient to internal and external factors of its production. Medieval authors also used the terms *coveniens* / *convenientia* and *congruum* / *congruentia*, close in meaning to *muṭābaqa*.

> Mais la pensée comme la parole doit se mesurer comme il faut à la nature du sujet: c'est ce qu'Albéric appelle très probablement la *congruentia*.

81 Bencheikh (1989), x.

82 Ṭaha Ḥusayn in the Egyptian television program "Najmuka al-mufaḍḍal" (your favourite star) (n.d).

> Beaucoup d'auteurs postérieurs insistent sur le principe essentiellement classique de *decorum* ou d'adéquation de l'expression au contenu.[83]

The meaning of *muṭābaqa* in Arab eloquence as attested in the *Muqaddima* confirms its vocation for designating rules of precision which do not admit approximation. But in this case Ibn Khaldūn does not refer to a rational method which guarantees the accuracy of *muṭābaqa*, as was the case in the field of historiography. Instead, he refers to the word "taste" (*dhawq*), a flair that guides the human being in the practice of arts and crafts and results from a habit (*malaka*) which becomes like a second nature for them.

> اعلم أن لفظة الذوق يتداولها المعتنون بفنون البيان و معناها حصول ملكة البلاغة للسان و قد مر تفسير البلاغة و أنها مطابقة الكلام للمعنى من جميع وجوهه بخواص تقع للتراكيب في إفادة ذلك فالمتكلم بلسان العرب و البليغ فيه يتحرى الهيئة المفيدة لذلك على أساليب العرب و أنحاء مخاطباتهم و ينظم الكلام على ذلك الوجه جهده فإذا اتصلت معاناته لذلك بمخالطة كلام العرب حصلت له الملكة في نظم الكلام على ذلك الوجه و سهل عليه أمر التركيب حتى لا يكاد يخطئ فيه عن منحى البلاغة التي للعرب و إن سمع تركيباً غير جار على ذلك المنحى مجّه و نبا عنه سمعه بأدنى فكر بل و بغير فكر إلا بما استفاده من حصول هذه الملكة فإن الملكات إذا استقرت و رسخت في محالها ظهرت كأنها طبيعة و جبلة لذلك المحل و لذلك يظن كثير من المغفلين ممن لا يعرف شأن الملكات أن الصواب للعرب في لغتهم إعرابا و بلاغة أمر طبيعي[84]

> You should know that the word "taste" [*dhawq*] is currently used by those who are concerned with the different branches of literary rhetoric [*bayān*]. "Taste" means that someone has acquired the habit of eloquence [*balāgha*]. We have already explained what "eloquence" means. It is conformity between words and meanings in every aspect and it is realised through special characters acquired by the words' combinations.
>
> The eloquent Arab speaker chooses the pertinent form [*ha'ya*] in this respect, according to the Arab models [*asālīb al-'Arab*] and the style of the Arab address. He makes every endeavour to compose the speech

---

83 De Bruyne (1946), II, 10–11, quotes Alberic of Montecassino (eleventh century).

84 Ibn Khaldūn (1970), III, 312–313.

> in this way. He attempts to make the best use of the Arab speech and in that manner he acquires the habit of composing the speech. Speech composition becomes so easy for him that he hardly fails to respect the rules of Arab eloquence. If he hears a combination of words that does not conform with the Arab path, he dismisses it and his ear rejects it without giving it any consideration, by virtue of the benefits he has drawn from that habit. When the habits become steadily established and rooted in their places, they are like a second nature, as if they were innate in those places. Many of those who ignore what habits are, think that the correct use of desinence and eloquence by the Arabs is natural.[85]

Ibn Khaldūn's use of *muṭābaqa* in the field of eloquence appears to be the result of intersections between Arab linguistics and logic. As far as logic is concerned Ibn Khaldūn probably relies upon Avicenna's writings, which he often quotes. There are two occurrences of *muṭābaqa* in Avicenna's logic which help us understanding the meaning of *muṭābaqa* in Khaldūn's linguistics. The first is to be found in Avicenna's definition of syllogism, the second in Avicenna's commentary on Aristotle's *Poetics*.

In the section on Logic of the *Najāt*[86] Avicenna, when he exposes syllogism, says that a "proposition" (*qaḍiyya*) is true if it necessarily and always "coincides" with "being" (*wujūd*). This is the case of the "universals" (*kulliyāt*). The "premise" (*muqaddima*) of a "syllogism" (*qiyās*) is true when its proposition about two entities A and B is always true, like the proposition "every moving being is a body" (*kull mutaḥarrik jism*),[87] which indeed coincides with being. Truth results from coincidence (*al-ṣidq huwwa bi al-muṭābaqa*).

> فإن الصدق هو بالمطابقة—وهذه المطابقة لا تحقق إلا فيما يجب الدوام له بل نحن لا نحكم في قضية محمولها ممكن وزمانها مستقبل بأنها صادقة أو كاذبة ما لم تطابق الوجود ولم تخالفه[88]

> Truth results from coincidence. Coincidence occurs only in what lasts. We can say that a proposition about something that can possibly happen in

85 Cfr. Ibn Khaldūn (1980), III, 358–359.

86 The *Najāt* [*Deliverance*] is a summary of the *Shifā'* [*Healing*, a monumental exposition of Aristotelian philosophy, but does not include *Poetics*.

87 Avicenna [e], *Najāt* (1985), 62.

88 Avicenna [e], *Najāt* (1985), 62.

> the future is true or false in as much as it coincides or does not coincides with being.[89]

The term *muṭābaqa* also occurs in Avicenna's commentary on the *Poetics* of Aristotle. The Arab commentators of Aristotle, following their Hellenistic sources, considered Aristotle's books on Rhetoric and Poetics as the last two books of Aristotle's writings on Logic. So does Ibn Khaldūn in the *Muqaddima.*[90] In Avicenna's commentary on Aristotle's Poetics *muṭābaqa* refers to a specific sort of similitude (*tashbīh*) that does not aim at praising or scoffing, but at a mere "equivalence" (*muṭābaqa*) between the intended meaning and its poetic representation.

و قد كان من الشعراء من يقصد التشبيه للفعل و إن لم يُخيل منه قبيحاً و حسناً، بل المطابقة فقط[91]

> Yet among the Greek poets there were also those who aimed with the comparison at the action itself, without creatively representing its ugliness or beauty, but only a simple equivalence (*muṭābaqa*)[92]

In conclusion of this section we can say that Ibn Khaldūn's historiographic notion of *muṭābaqa* is full of echoes from other contexts and disciplines in which this term occurs in classical Arab culture, and particularly linguistics and logic. The intertextuality of this term shows the *systematic* character of the *Muqaddima*, its being a picture of the Arab-Islamic culture in the classical era.

3 Parallel Developments of the Root "ṭ b q" (*inṭibāq*, *taṭbīq*) in Arab Euclidean Geometry

In this section we would like to draw the attention of the reader on parallel developments of the root "ṭ b q" in the field of Arab Euclidean geometry. It is worthwhile mentioning that the root "ṭ b q" had parallel developments in Arab translations and commentaries on Euclid's *Elements*, where *inṭibāq* and *taṭbīq*

89 Avicenna [e], (1985), 62.

90 Ibn Khaldūn (1970), III, 108f., where he quotes al-Fārābī, Avicenna and Averroes' commentaries on Aristotle's poetics. Cfr. Ibn Khaldūn (1980), III, 147–147.

91 Avicenna [f], *Shi'r* [*Poetics*] (1966), 30.

92 Vincente Cantarino, *Arab Poetics in the Golden Age* (1975), 139.

occur in the axiom on equality of "superimposable" figures. We argue that these developments show analogies with the notion of *muṭābaqa* in Khaldūn's historiography due to their common logic background.

The *Elements* (*Stoichéia*) by Euclid (fl. third century BCE) begins with a series of non-demonstrated principles, which are the fundament of the *Elements*' exposition on geometry. These principles are the definitions (*óroi*), the postulates (*aithḗmata*) and the axioms or common notions (*koinái énnoia*). Fabio Acerbi in his edition and translation of the *Elements* observes that Aristotle mentions a similar division in the *Posterior Analytics* (part of his writings on Logic).[93] In the Arab translation of the latter by Abū Bishr Mattā ibn Yūnus (870–940) realised from a Syriac translation by Ḥunayn Ibn Isḥāq (808–873), we read that "the axioms are primary truths from which demonstration departs."

العلوم المتغارفة التي يقال لها عامية و هذه هي اللأوائل منها يبينون

> Les principes communs, appelés axiomes, vérités premières d'après lesquelles s'enchaine la démonstration[94]

The "common notions" or "axioms" of Aristotle's logic acquired a technical and theoretical meaning in Euclid's *Elements* and their Arab versions. In the standard Arab translation of the *Elements* by Ḥunayn Ibn Isḥāq revised by Thābit Ibn Qurra (836–901) they are called "*al-ʿulūm al-mutaʿārifa*"[95] (recognised notions), although "*ʿilm ʿāmm muttafiq ʿalayhi*"[96] (general notions agreed upon) also occurs. Avicenna in his *Uṣūl al-handasa* (Fundaments of Geometry) instead says "*ʿilm jāmiʿ*" (general notions). Among the axioms, there is the one stating that figures that can be "superimposed" on one another are equal.

و ما انطبق على آخر انطباقاً لا يفضل أحدهما على آخر مساو له[97]

> Things that can be superimposed one upon another without exceeding one another are equal to each other

93 Euclid, *Elements* (2007–2014), 218.
94 Rashed, *Lexique historique de la langue scientifique arabe* (2017), 563.
95 Quoted by Rashed (2017), 564.
96 Rashed (2017), 564.
97 Avicenna [g], *Uṣūl al-handasa* [Principles of Geometry] (2012), 19–20.

Ibn Khaldūn uses the term *inṭibāq* as quasi synonymous of *muṭābaqa* in two relevant cases: a) when he refers to a mental form (*ṣūra dhihniyya*) conforming with particular word combinations (*tarkīb khāṣṣ*) in the act of poetical composition;[98] b) when he refers to conformity between the first intelligibles (*al-maʿqūlāt al-uwwal*) and external reality (*mā fī al-khārij*) in the practice of logic.[99]

We have already mentioned Ibn Khaldūn's appreciation for Ibn al-Haytham's optics, a branch of geometry. Shortly before he briefly expresses his appreciation for the geometrical sciences in general, for their coherence (*intiẓām*), the order (*tartīb*) of their proofs (*barāhīn*) and the impossibility of error in them.

و اعلم أن الهندسة تفيد صاحبها إضاءة في عقله استقامة في فكره لأن براهينها كلها بينة الانتظام
جلية الترتيب لا يكاد الغلط يدخل أقيستها لترتيبها و انتظامها فيبعد الفكر بممارستها عن الخطاء
و ينشأ لصاحبها عقل على ذلك المهيع[100]

> You should know that geometry is useful to those who attend to it: it enlightens their intellect and makes their intelligence right because all its logical proofs are clearly ordered so that it is almost impossible that errors penetrate them. The mind of those who practice geometry is far from error and and their intellect grows when they follow this path.[101]

We have no evidence of a direct impact between Euclidean geometry and Khaldūn's historiography, but we remark an analogy between them. The Euclidean axiom states that superimposable figures are equal. Khaldūn's *axiom* in historiography states that entities that are superimposable are real and true. This analogy is not surprising because it derives from interactions within the *Muqaddima*-system.

4 The Nature of Civilisation (*ṭabīʿat al-ʿumrān*)

In previous sections of this article we have seen that Ibn Khaldūn reinterprets philosophical Aristotelian categories in order to develop his own materialistic and mechanistic conception of history. Particularly, we have seen that he turns the Avicennian psychological notion of *muṭābaqa* into an immanent law of

98 Lelli (2014), 210.
99 Cfr. *supra* in section four: "*Muṭābaqa* and philosophy".
100 Ibn Khaldūn (1970), III, 102.
101 Cfr. Ibn Khaldūn (1980), III, 130–131.

history. In this section we analyse how Ibn Khaldūn applies to the "nature of civilisation", categories of the Aristotelian physics, describing change in terms of matter and form, power and act. This analysis allows us to highlight further that the *Muqaddima* is a book of intersections between classical Arab thought and a new conception of history, and between scientific and social thought.

The very purpose of the *Muqaddima* is the study of conditions affecting the "nature of civilisation" (*ṭabīʿat al-ʿumrān*).[102] Civilisation belongs to the physical worlds, the world of changing entities. Khaldūn insists that the historian must take into consideration "change in conditions" (*tabaddul aḥwāl*) in the *longue durée* (*ayyām, zaman*). He warns that passed conditions cannot be assimilated critically by analogy to new ones. Therefore, as we have seen above (section 3), "analogical reasoning" (*qiyās*) should be applied scientifically, critically and not mechanically.

According to Ibn Khaldūn's well known organicist conception of history, civilisation is a living being. Therefore, it is born, grows, enter into decay and dies. Mineral, plants, animals, dynasties, conditions, sciences and arts are all physical entities that come into being, grow and enter into decay according to the same laws of nature.

> اعلم أن العالم العنصري بما فيه كائن فاسد لا من ذواته و لا من أحواله و المكونات من المعدن و النبات و جميع الحيوانات الإنسان و غيره كائنة فاسدة بالمعاينة و كذلك ما يعرض لها من الأحوال و خصوصا الإنسانية فالعلوم تنشأ ثم تدرس و كذلك الصنائع و أمثالها[103]

> You should know that the world of the elements and all its beings are corruptible. This applies to its essences and conditions, to the minerals, the vegetables and the animals, including the human beings and all other beings which are all corruptible, as it can be observed. The conditions that affect [all these beings] are also corruptible, especially human conditions. The same applies to the sciences, the arts and similar things that grow then are wiped out.[104]

---

102 The subject of book one of the *Kitāb al-ʿIbar* (*Book of Lessons*) is indeed history, which "in matter of fact, is information about human society, that is to say word's civilisation, and about conditions that affect it as, for example, savagery and sociability, group feeling and the different manners in which a group of human beings turn against another group", cfr. Ibn Khaldūn (1980), I, 69. لما كانت حقيقة التاريخ أنه خبر عن الاجتماع الانساني الذي هو عمران العالم و ما يعرض لطبيعة ذلك العمران من الأحوال مثل التوحش و التأنس و العصبيات و أصناف التقلّبات للبشر بعضهم على بعض, Ibn Khaldūn (1970), I, 56.

103 Ibn Khaldūn (1970), I, 247–248.

104 Cfr. Ibn Khaldūn (1980), I, 278.

Ibn Khaldūn describes becoming and change in history with the Aristotelian terms of matter and form, power and act. Civilisation (*'umrān*) is matter. States (or dynasties) are its form (*ṣūra*). Primitive Bedouin civilisation naturally acquires the form of the State (*dawla*). "Matter" and "form" acquire a concrete social and economic meaning, according to Ibn Khaldūn's materialistic conception of history. The nature of civilisation is political power ("the different manners in which one group of human beings imposes its superiority upon over another"). The "matter" of civilisation are the economic resources accumulated by States, which detain the political authority (*mulk*). The States and the rulers are the world's biggest market.

Curtailment of the allowances given by the State provokes curtailment of tax revenue and a chain reaction leading to the fall of demand and irrecoverable impoverishment of the society. Ibn Khaldūn describes the economic system contemporary to him in terms of tributary mode of production (although he does not use this term), in as much as the surplus necessary to its reproduction derives from taxes collected by the State.[105]

> الدولة والملك للعمران بمثابة الصورة للمادة و هو الشكل الحافظ بنوعه لوجودها و قد تقرر في علوم الحكمة أنه لا يمكن انفكاك أحدِهما عن الآخر فالدولة دون العمران لا تتصور و العمران دون الدولة و الملك متعذر بها في طباع البشر من التعاون الداعي إلى الوازع فتتعين السياسة لذلك أما الشريعة أو الملكية و هي معنى الدولة وإذا كانا لا ينفكان فاختلال أحدهما مؤثر في اختلال الآخر كما كان عَدَمه مؤثّراً في عدمِه[106]

> The State and political authority have the same relationship to civilisation as form to matter. The form is the shape that preserves the existence [of matter] through the [particular] kind [of phenomena that affects it]. Philosophy has established that it is not possible to separate the one from the other because a State without civilisation is inconceivable and civilisation without State and political authority is impossible: the human beings are naturally disposed to cooperation and [cooperation] needs a restraining influence.

105 For a definition of the tributary mode of production: "le mode de production 'tributaire', qui adjoint à la persistance de la communauté villageoise un appareil social et politique d'exploitation de celle-ci sous la forme de la ponction d'un tribut; ce mode de production est la forme la plus courante qui caractérise les formations de classes précapitalistes", Samir, *Le développement inégal* (1973), 9–48.

106 Ibn Khaldūn (1970), II, 264–265.

> The exercise of power, be it religious or political, needs [that restraining influence], which is the State. Given that [State and political authority] are inseparable, a deficiency affecting one of them affects the other and if one of them does not exist, the other does not exist either.[107]

In section two we have proposed a distinction between two levels in Ibn Khaldūn's exposition on history. The first level is the level of micro dimensions, *courte durée* and single events. To this level Khaldūn applies explicitly the law of *muṭābaqa*, which is affected by finalism and apriorism because micro events happen as a consequence of natural characters inherent in their matter. Happening events realise in act "the possibility [*imkān*] of the matter that belongs to a given thing". "Possibility" should not be intended in a general sense of an option among others, but in the narrow sense of the potentiality inherent in a certain matter characterised by certain genus, difference, size and strength. Possibility is therefore very close to necessity. This level reflects the Aristotelian conception of change and remains affected by finalism and apriorism.

> وهذا كثيراً ما يعتري الناس في الأخبار كما يعتريهم الوسواس في الزيادة عند قصد الإغراب كما قدمناه أول الكتاب فليرجع الإنسان إلى أصوله و ليكن مهيمناً على نفسه و مميزاً بين طبيعة الممكن و الممتنع بصريح عقله و مستقيم فطرته فما دخل في نطاق الإمكان قبله و ما خرج عنه رفضه و ليس مرادنا الإمكان العقلي المطلق فإن نطاقه أوسع شيء فلا يفرض حداً بين الواقعات و إنما مرادنا الإمكان بحسب المادة التي للشيء فإنا إذا نظرنا أصل الشيء و جنسه و فصله و مقدار عظمه و قوته أجرينا الحكم في نسبة ذلك على أحواله و حكمنا بالامتناع على ما خرج من نطاقه[108]

> With regard to historical information, people are often tempted to exaggerate with the purpose of reporting something amazing, as we mentioned at the beginning of this book. Therefore, one must refer to one's sources and [think] independently. He must distinguish with a clear mind and a sound instinct between what is naturally possible and what is naturally impossible. What falls within the domain of the possible should be accepted. What falls within the domain of the impossible should be rejected. By "possible" we do not mean what is intellectually

---

107 Cfr. Ibn Khaldūn (1980), II, 300–301.
108 Ibn Khaldūn, (1970), I, 329.

possible in an absolute sense, because its domain is too large and historical facts cannot be precisely delimited within it. By "possible" we mean what could be possible if we take into account the matter that pertains to a certain thing. When we carefully consider the essence of a thing, its genus, property, size and power, we can judge its conditions on those grounds and conclude that all that falls outside this domain is impossible.[109]

The second level is the level of macro dimensions, *longue durée* historical processes mainly determined by economic factors and described in a mechanistic but no longer, or not exclusively, naturalistic way. This level is closer to Karl Marx's materialistic description of society than to Aristotle's philosophy and describes the functioning of the tributary mode of production that undergoes cyclical crisis of underproduction and decay.

فصل في أن نقص العطاء من السلطان نقص في الجباية
و السبب في ذلك أن الدولة و السلطان هي السوق الأعظم للعالم و منه مادة العمران فإذا احتجن السلطان الأموال و الجبايات أو فقدت فلم يصرفها قل حينئذ ما بأيدي الحاشية و انقطع أيضا ما كان يصل منهم لحاشيتهم و ذويهم و قلت نفقاتهم جملة و هم معظم السواد و نفقاتهم أكثر مادة للأسواق من سواهم فيقع الكساد حينئذ في الأسواق و تضعف الأرباح في المتاجر لقلة الأموال فيقل الخراج لذلك لأن الخراج و الجباية إنما يكون من الاعتمار و المعاملات و نفاق الأسواق و طلب الناس للفوائد و الأرباح و وبال ذلك عائد على الدولة بالنقص لقلة أموال السلطان حينئذ بقلة الخراج فإن الدولة كما قلناه هي السوق الأعظم أم الأسواق كلها و أصلها و مادتها في الدخل و الخرج فإن كسدت و قلت مصارفها فاجدر بما بعدها من الأسواق أن يلحقها مثل ذلك و أشد منه و أيضا فالمال إنما هو متردد بين الرعية و السلطان منهم إليه و منه إليهم فإذا حبسه السلطان عنده فقدته الرعية[110]

A diminution of gifts bestowed by the ruler means decrease of tax revenue
The reason for that is that the State and the ruler are the world's biggest market from which the substance [*mādda*] of civilisation originates. If the ruler takes hold of capital and tax revenue, or if capital and tax revenue

109 Cfr. Ibn Khaldūn (1980), I, 371–372.
110 Ibn Khaldūn (1970), II, 92–93.

> are lost and therefore they are not spent, then the wealth at the disposal of his entourage also diminishes and their own entourage and relatives stop receiving a part of the wealth in their turn and their expenditures diminish greatly. The latter represents the majority of the people and their expenditures provide more substance [*mādda*] to the market than any other group of people. As a consequence of that, the profit deriving from trade declines because of a lack of capital. The land tax revenue diminishes because land tax and other tributes depends on prosperity, on transactions, on capital spent in the markets and on the people's demand for gain and profit. This situation has an impact on the State, which is affected by a deficit, due to a lack of capital at the disposal of the ruler, as a consequence of the decrease of the revenue deriving from the land tax.
>
> As we have said, the State is the biggest market, the market of all markets. The State is the origin and the substance of all markets with regard to income and expenditure. If the State stagnates and the volume of economic activities diminishes, the other markets depending on it will follow and they will be even more seriously affected. Wealth circulates between the ruler and the subjects, but if the ruler keeps it to himself, the subjects lose it.[111]

Our modern judgment perceives an important qualitative difference between these two levels: the micro level of *courte durée* is Aristotelian, the macro level of *longue durée* is historical-materialist.[112] Apparently this gap does not preoccupy Ibn Khaldūn. As a good medieval scholar, he wanted to integrate all possible features of his culture in a "system". What Ibn Khaldūn calls the "nature of civilisation" is a notion in the intersection between different disciplines and even between different worlds. It lies between Aristotelian physics and Ibn Khaldūn's new materialistic conception of history. It lies also between scientific and social thought in the modern sense.

Marx has observed that before Newton and before his and Engels' materialistic conception of history, materialism in European thought was a field of intersections between the science of nature and the sciences of man. Trends of European thought in the seventeenth century ascribed to matter a self-creative force. Others wanted to apply the rules of geometry to society and human intellect.[113] Ibn Khaldūn, who still lived in an ancient world, was already operating

111 Ibn Khaldūn (1980), II, 102–103.

112 For Karl Marx's systematic materialistic conception of history cfr. Marx's *Critique of Political Economy* (1859), Marx (1993).

113 Considerations on philosophical materialism as a phase of transition and intersections

in these intersections, very much in advance with respect to the conditions around him.

5 Conclusion

The general purpose of this article was to show that scientific thought, like all other aspects of human thought, developed thanks to exchanges and intersections between different fields and disciplines. This phaenomenon has been particularly evident in the European thought of the seventeenth and eighteenth centuries, when philosophical materialism did not distinguish clearly between the laws of nature and the laws of man and society. Similar intersections can be found in Ibn Khaldūn's philosophy of history exposed in the *Muqaddima*. His materialistic and mechanistic conception of history indeed lies in the intersection between classical Islam and modern materialism, between social and scientific thought.

In this article we have analysed specifically the notion of *muṭābaqa* ("coincidence", "conformity") in Ibn Khaldūn's philosophy of history. If the *Muqaddima* is a *system* that includes to the maximum possible extent all the components of classical Arab thought, the *muṭābaqa* is a microcosm of this system and contains much of the *Muqaddima* itself.

We have seen that according to Ibn Khaldūn *muṭābaqa* is an objective law of history that applies namely to micro dimensions, at the *courte durée* and single event level. It means "coincidence" between events (*wāqiʿ, ḥādith*) and conditions (*aḥwāl*). *Muṭābaqa* is also an instrument at the disposal of the historian for verifying this coincidence in historical sources with the precision of a geometer. Moreover, *muṭābaqa* is an epistemological criterion of truth: coincidence between the psychological representation of history by the historian and external historical reality (*wujūd*) guarantees certitude of knowledge.

We have highlighted that Ibn Khaldūn's historiographic *muṭābaqa* is affected by a finalistic and aprioristic aporia (contradiction), and we have analysed the content and the reasons for this aporia. As for its content, we have seen that according to Ibn Khaldūn historical events happen in virtue of pre-existing attributes inherent in their material nature. Once they have become actual, events necessarily "coincide" with "conditions". The historian attains a perfect

---

between materialism of nature and materialism of society have been expressed by Engels-Marx, *The Holy Family* (1845), passages quoted in Karl Marx and Friedrich Engels, *La concezione materialistica della storia*, ed. Nicolao Merker (1998), 41–49.

knowledge of events when his mental representation of them (the first intelligibles) "coincide" with their external sensual reality. We have identified two main reasons for this aporia:

a. Ibn Khaldūn's materialism does not contemplate the power of human actors to shape historical conditions and determine the course of history, as Karl Marx and non-dogmatic Marxists would do.
b. Ibn Khaldūn ascribes to the law of *muṭābaqa* a precision impossible to attain in social sciences. Only modern physics would be able to formulate general laws of nature expressed with mathematical formula and experimentally verifiable.

Ibn Khaldūn relies upon Avicenna's Psychology, a branch of Aristotelian physics which, since Aristotle himself, was affected by Platonism. In formulating the law of *muṭābaqa*, and in order to give coherence to his materialistic conception of history, Ibn Khaldūn conceives an immanent kind of Neo-Platonism brought down to earth, but not delivered from apriorism and finalism.

Italian Humanists of the fifteenth century used Neo-Platonism in order to find new human dignity and freedom, supported by emergent capitalism and its new urban classes proud of their initiative and creativity. This was not the case for Ibn Khaldūn, who was surrounded by social decay and human distrust. Nevertheless, Ibn Khaldūn's conception of history is not pure and cold mechanism. It is a humanism in reverse. Ibn Khaldūn's humanism consists precisely in his pessimistic and tragic awareness of the impotence of man in the course of history.

Bibliography

ʿĀbid al-Jabrī, Muḥammad [currently mentioned as Abed al-Jabri, Mohammed], *Naḥnu wa al-turāth* [*We and our Heritage*] (Beirut: Markaz dirāsāt al-waḥda al-ʿarabiyya, 2006).

Amin, Samir, *Le développement inégal* (Paris: Editions de Minuit, 1973).

Aristotle (384–322 BCE) [a], Poetics [*perì tekhnēs poiētikēs*], transl. Diego Lanza, *Poetica* (Milano: Biblioteca Universale Rizzoli, 1990), Greek text same as *Poetics* (1965).

Aristotle [b], *Poetics*, ed. R. Kassel, *Aristotelis De Arte Poetica Liber* (Oxford: Clarendon Press, 1965).

Aristotle [c], *Poetics*, transl. Samuel Henry Butcher, *The Poetics of Aristotle* (London: Macmillan, 1902), https://www.stmarys-ca.edu/sites/default/files/attachments/files/Poetics.pdf (URL accessed 13 April 2021).

Aristotle, [d], *The Soul* [*Perì psychēs* / *De Anima*], transl. Giancarlo Movia, *L'Anima* (Milano: Rusconi, 1996).

Aristotle, [e], *The Soul* [*Perì psychēs / De Anima*], ed. David Ross (Oxford: Clarendon Press, 1961).

Avicenna [a][Ibn Sīnā] (980–1037), *Psychology* [*ʿIlm al-nafs*], ed. Ján Bakoš, *Al-fann al-sādis min al-ṭabīʿiyyāt (ʿilm al-nafs) min Kitāb al-Shifāʾ taʾlīf Ibn Sīnā*, I, (Prague: Editions de l'Académie Tchécoslovaque des Sciences, 1956).

Avicenna [b], *Psychology* [*ʿIlm al-nafs*], transl. Ján Bakoš, *Psychologie d'Ibn Sīnā (Avicenne) d'après son oeuvre ash-Shifāʾ*, II (Prague: Editions de l'Académie Tchécoslovaque des Sciences, 1956).

Avicenna [c], *The book of Healing* [*Kitāb al-Shifāʾ*] (Lithographed in Teheran, 1886) [2 vols.]).

Avicenna [d], *The book of Deliverance* [*Kitāb al-Najāt*] (Cairo, 1913).

Avicenna [e], *The book of Deliverance*, ed. by Mājid Fakhrī, *Kitāb al-Najāt* (Beirut: Manshūrāt al-afāq al-jadīda, 1985).

Avicenna [f], *Poetics*, ed. by ʿAbd al-Raḥmān Badawī, *Kitāb al-Shiʿr* (Cairo, 1966).

Avicenna [g], *Principles of Geometry* [*Uṣūl al-handasa*]: *Al-Shifāʾ. Al-fann al-awwal min jumla al-ʿilm al-ryāḍī. Uṣūl al-handasa*, ed. A. Sabra and A. Lotfi, (Teheran: s.n., 2012).

Bencheikh, Jamel Eddine, *Poétique arabe* (Paris: Gallimard, 1989).

Braudel, Fernand, "History and the Social Sciences. The Longue Durée", transl. Immanuel Wallerstein, *Review* XXXII, 2 (2009), 171–203. First published in French in 1958.

Cantarino, Vincente, *Arab Poetics in the Golden Age* (Leiden: Brill, 1975).

De Bruyne, Edgar, *Etudes d'Esthétique médiévale*, Vols. I–III (Brugge: De Temple: 1946).

Descartes, René, *Principia Philosophiae*, transl. *Valentine Rodger and Reese P. Miller (Dordrecht: Reidel, 1983), 1st Latin edition* 1644.

Eco, Umberto, *Le problème esthétique en Thomas d'Aquin* (Paris: Presses Universitaires de France, 1993).

Eco, Umberto, *Il problema estetico in Tommaso d'Aquino* (Milano: Bompiani, 1982).

Abdelali Elamrani, Jamal, "Prophétie selon Ibn Khaldūn et philosophie arabe classique", in *Ibn Khaldūn et la foundation des sciences sociales*, ed. Zeïneb ben Saïd and Georges Labica (Paris: Publisud, 2009).

Euclid (fl. 300 BCE), *Elements*, in *Euclide. Tutte le Opere*, ed. and transl. Fabio Acerbi (Milano: Bompiani, 2007–2014).

Gabrieli, Francesco, "Ibn Khaldūn. Il Vico dell'Islam", *Bollettino del Centro di studi vichiani*, V (1975), 122–126.

Gabrieli, Francesco, *La letteratura araba* (Firenze-Milano: Sansoni-Accademia, 1967).

Goichon, Amélie, *Lexique de la langue philosophique d'Ibn Sina* (Paris: Desclée de Brower, 1938).

Hobbes, Thomas, *Leviathan*, ed. J.C.A. Gaskin (Oxford: Oxford University Press, 2008), 1st edition 1651.

Hume, David (1711–1776), *A Treatise of Human Nature*, ed. David Fate Norton and Mary J. Norton (Oxford: Oxford University Press, 2007), 1st edition 1739–1740.

Ḥusayn, Ṭaha, intervention in the Egyptian television program "Najmuka al-mufaḍḍal" نجمك المفضل: شمس المعرفة طه حسين وسط نجوم الفكر والأدب (n.d), URL https://www.youtube.com/watch?v=gK-e4LUFtok (URL accessed 12 April 2021).

Ibn al-Haytham (965–1040) [a], *Kitāb al-manāẓir*, ed. Abdelḥamīd I. Ṣabra, *Edition of the Arabic Text of Books I–III: On Direct Vision* (Kuwait: National Council for Culture, Arts and Letters, 1983).

Ibn al-Haytham [b], *Kitāb al-manāẓir*, ed. Abdelḥamīd I. Ṣabra, *The Optics of Ibn al-Haytham, Edition of the Arabic Text of Books IV–V: On Reflection and Images Seen by Reflection, Vols. I–II.* (Kuwait: The National Council for Culture, Arts and Letters, 2002).

Ibn al-Haytham [c], *Kitāb al-manāẓir*, transl. Abdelḥamīd I. Ṣabra, *The Optics of Ibn al-Haytham. Books I–II–III: On Direct Vision, Vols. I–II* (London: The Warburg Institute, 1989).

Ibn Khaldūn, *Muqaddima*, transl. William Mac-Guckin De Slane, *Prolègomènes historiques d'Ibn Khaldūn*, in *Notices et extraits des manuscrits de la Bibliothèque Impériale* (Paris: 1862, 1865 and 1868 [vols. XIX–XXI]).

Ibn Khaldūn, *Muqaddima*, transl. Franz Rosenthal, *Muqaddima. An Introduction to History*, Vols. I–III (Princeton: Princeton University Press, 1980).

Ibn Khaldūn, *Muqaddima*, ed. Etienne Quatremère, Prolégomènes d'Ebn *Khaldoun*, Vols I–III (Paris: 1958, Reprint Beirut: Librairie du Liban, (1970).

Ifrah, Gorges, *Histoire universelle des chiffres* (Paris: Editions Robert Laffont, 1994).

Jolivet, Jean, "Classification of the Sciences", in *Encyclopedia of the History of Arabic Science*, ed. Rosdhi Rashed, Vols. I–III (London and New York: Routledge, 1996), III, 1008–1025.

Khalīl Ibn Aḥmad, *Kitāb al-ʿAyn*, ed. Mahdī al-Makhzūmī and Ibrāhīm al-Sāmarrāʾī (Beirut: Dār wa-Maktabat al-Hilāl, 1986).

Khalīl Ibn Aḥmad (721–791) [b], *Kitāb al-ʿAyn*, ed. Mahdī al-Makhzūmī and Ibrāhīm al-Sāmarrāʾī, Vols. I–VIII (n.l.: s.n., n.d.) Vol. V, URL https://archive.org/details/alaeen_Farahidi (URL accessed 12 April 2021). Likely the same edition as Khalīl Ibn Aḥmad (1986).

Koyré, Alexandre, *Du monde de l'à peu près à l'univers de la précision*, in A. Koyré, *Etudes d'histoire de la pensée philosophique* (Paris: Gallimard, 1948).

Lelli, Giovanna, "Neo-platonism in Ibn Khaldūn's Poetics", *Al-Masāq. Journal of the Medieval Mediterranean*, 26:2 (2014), 196–215.

Locke, John (1632–1704), *An Essay Concerning Human Understanding*, ed. Peter H. Nidditch (Oxford: Oxford University Press, 1975), 1st edition 1689.

Marx, Karl (1818–1883), *Per la critica dell'economia politica*, transl. E. Cantimori Mezzomonti (Torino: Editori Riuniti, 1993).

Marx, Karl and Engels, Friedrich, *La concezione materialistica della storia*, ed. by Nicolao Merker (Torino: Editori Riuniti, 1998).

Pellat, Charles, "Adab ii. Adab in Arabic Literature," *Encyclopædia Iranica*, I: 4 (1983), 439–444, URL http://www.iranicaonline.org/articles/adab-ii-arabic-lit (URL accessed 12 April 2021).

Rashed, Roshdi, "Algèbre et linguistique. Le début de l'analyse combinatoire", in *D'al-Khwārizmī à Descartes. Études sur l'histoire des mathématiques classiques* (Paris: Hermann, 2011).

Rashed, Roshdi (ed.), *Lexique historique de la langue scientifique arabe* (Hildesheim: Georg-Olms-Verlag, 2017).

Sammarchi, Eleonora, "L'algèbre des polynômes. Le Qisṭās al-muʿādala fī ʿilm al-jabr wa ʾl-muqābala d'al-Zanjānī (XIIIe siècle)" PhD Dissertations, Vols. I–II (Université Sorbonne Paris Cité, Préparée à l'Université Paris Diderot, 2017).

Van Dalen, Benno, "Battānī: Abū ʿAbd Allāh Muḥammad ibn Jābir ibn Sinān al-Battānī al-Ḥarrānī al-Ṣābiʾ", *The Biographical Encyclopedia of Astronomers* ed. T. Hockey et al. (New York: Springer, 2007), 101–103.

T.V. Venkateswaran, "Influence of Ibn al-Haytham on Vision, Optics and Experimental Sciences", *Science and Culture* 82: 11–12 (2016), 358–363.

CHAPTER 10

Arabic Medicine in China: Context and Content

Paul D. Buell

Cinggis-qan (d. 1227) and their successors created the largest empire in history, and although the Mongol hordes have been most famous for rapine, pillage, war, and conquest, their overall reputation has recently achieved a well-deserved and long-awaited rehabilitation, based on Mongol achievements in many other areas than empire building.[1] A new generation of scholars (led by Jack Weatherford) now recognises that the Mongols, when they were not conquering and setting up empires and states, were often busy spreading cultural, technological and even scientific goods from one part of the world to the other, everything from food to philosophy and medicinals and medical lore, as well as achievements of science and technology.

Among transfers to the West was printing, including the first European paper money (already issued at the end of the fourteenth century),[2] guns and gunpowder,[3] many innovations in fabrics and fabric-making, included printed fabrics, although the Iranians had that technology too, ceramics and the how-to that ultimately led up to the first European experiments with porcelain-making,[4] bowed-fiddling,[5] probably the whole idea of the lute song with lutes strummed like Central Asian instruments, and not plucked like Arabic lutes, probably a great deal of maritime technology (fore-and-aft rigging and the stern-post rudder, certainly the compass, although some of the transfers had

---

1 The term Arabic medicine is used in this paper because the Arabic language was the major vehicle for its transmission and even for Persian materials it was the primary source of technical terminology. See also the discussions in Buell and Anderson (2021).

2 On paper money in China see now Vogel (2012). Polo's descriptions of paper money would, of course, have been read in Europe. Venice, from where he came, was among the first, in fact, to introduce paper money. It was still not printed but this innovation came soon.

3 On these transfers in general see the rich material in Needham, et al., (1954–2015).

4 For a general introduction of porcelain outside of China see Carswell (2000).

5 In the epic *Der Nibelungen Not* of 1200 ca., Volker, *der fidelaere*, the "fiddler", actually plays a cranked hurdy-gurdy to tell his tale of the Burgundians and Huns, but a 100 years or so later, while there were still a few hurdy-gurdy players around, the style of music had changed completely and violins, violas and other bowed string instruments were the rage. This was new for Europe, but old hat in the steppe, where horse-headed fiddles and its ancestors, and *kobyz*, the Kazakh violin, were a long tradition. The European change is suspicious and almost certainly is to be associated with Mongol influence.

 | DOI:10.1163/9789004513402_012

been going on before the Mongols), and warfare theory and technology. The Portuguese fought the decisive battle of Aljulbarrota (1385) entirely using Mongol strategy.[6] Even vodka distillation is now thought to have used technology that came in with the Mongols and folk stills in places as far afield as Iceland, where they were seen by the present author in a museum, look more like those popularised by the Mongols, even made of wood like theirs, than anything related to the glass and metal Mediterranean tradition with its serpentines. Transfers to the East included, among other things, art and painting techniques, the vibrant color of many Yuan 元 Dynasty paintings, for example, although this was also based in Tang 唐 tradition, and most of what was new was Iranian, foods, including Mongol court cuisine, also mostly Iranian, some of it filtered by Turkic traditions, philosophy and religion, and medicine, from all over Eurasia, in particular many new medicinals. The cobalt-blue pigment used in manufacturing Blue and White Porcelain was at least popularised in Mongol times. It was mostly imported. Maps and map-making came too[7] and astronomy, with tables purely from Iran but further developed in Mongol China.

The Mongols became a more dramatic force for "globalisation" than the namesake late twentieth century movement that has seen a great expansion of goods, and the cultures associated with them across the globe. Rap may be an American invention, but the present author has seen it performed in Mongolia, while Korean girl bands are all the rage in Seattle. McDonalds can be found in Ōsaka, while Japanese sushi is now everywhere in the US. And movement was rapid in Mongol times. The Mongol pony-express, the *jam*, mostly using horses, but even runners and dog sleds in the far north, shortened the time for a letter to go from Baghdad to Beijing from several months to a few days. In addition, as the Mongol conquest of China "took", the Mongols "took" to the seas in a big way, defeating Southern Song 宋 in a great sea battle in 1279, in its own element,[8] threatening its neighbors with seaborne invasions, even as far as Java, and when direct conquest failed, giving rise to an impressive ocean commerce, part of an enhanced Maritime silk road, one linking Iran and the rest of the Middle East, also Europe via the Black Sea,[9] with Zayton (Quanzhou 泉州) and other Chinese ports.

Primarily because of geography, Mongol preoccupation with China were of long standing. China was physically close and Chinese dynasties and states

6 See, as an introduction to the battle, Rosa (2008). See also, the archaeological accounts in Monteiro (2001).
7 See as an introduction Park (2012).
8 See Buell (1985–1986).
9 See now Ciolcîltan (2012).

used to meddle in the steppe when they did not invade if directly, when they were feeling particularly strong or forced to resist steppe raids and incursions into China itself. Among those working for the Jürched, and the North Chinese Jin 金 Dynasty (1125–1234), was Cinggis-qan himself, although this part of his biographical history is largely suppressed in the *Secret History of the Mongols*, more hagiographical in its approach than purely factual.

Also because of geography and history, the first Mongol attacks on the outside world were against China, in this case the territories of the Jin Dynasty and the Tangut Xixia 西夏 state (1038 to 1227). The Mongols at first raided, starting with Xixia in 1205, and after 1215 began conquering China piecemeal, although even before 1215 they had formed many alliances with local people that in effect obligated these local people to hold territory in their name and help the Mongols militarily and administratively.

Xixia was finally conquered in 1227, the year of Cinggis-qan's death, and Jin in 1234–1235. Even before that date raids and serious warfare had begun against Song 宋. The last *Qan* of unified Empire, Möngke (r. 1251–1259) died while campaigning there. His death launched a protracted series of civil wars. The Mongol unified empire broke up into competing regional units, in Russia, the Golden Horde, in Iran, the Ilqanate, the Ca'adai qanate in western Turkistan, and later another unit in the area representing the house of Ögödei (r. 1229–1241), deposed when Möngke came to power. In China Qubilai (r. 1260–1294) established his independent power base and founded qanate China. In 1271 this qanate became the Yuan 元 Dynasty. In 1279, after a protracted struggle, Qubilai's armies conquered Song, unifying China for the first time since the twelfth century. Qubilai's successors continued to rule until 1368, mostly with short reigns except the last emperor, Toγon-temür (r. 1332–1368) who had to abandon Qanbaliq and flee Ming troops. Ultimately the entire Dynasty moved back to Mongolia.[10]

During Qubilai's early reign and the Yuan proper, China was remarkably cosmopolitan. Not only was there the rule of a foreign dynasty, but Qubilai welcomed visitors and subjects from almost everywhere including, the most famous, Marco Polo (1254–1324). Also prominent at court and elsewhere were Turkic-speakers, some local, some from a great distance. They were particularly conspicuous in government including in positions of special expertise as court cooks and doctors. For one example, the Yuan court nutritional manual of 1330, *Yinshan zhengyao* 飲膳正要, "Proper and Essential Things for the Emperor's Food and Drink", contains substantial Turkic elements and

10 On the Mongols as an empire and afterwards see Buell and Fiaschetti (2018).

its author, Hu Sihui 忽思慧, was almost certainly Turkic in culture. Also conspicuous were Koreans—the last Mongol empress and regent was Korean, for example, Empress Ki (d. 1370 ca.), the survivors of local groups such as Tanguts and Jürched, and a great variety of persons with Middle Eastern connections, usually just through religion, e.g. Islam or Christianity, and mainly merchants, but also scientists and doctors. This was the period also when Tibetan Tantric Buddhism became the official religion (Tibetans also were involved in Medicine, including dietary medicine, before Hu Sihui, and in astronomy) and even Catholic Christianity got a start not equaled until the period after the sixteenth century. Ming 明, incidentally, continued with these traditions of cosmopolitanism, at least early Ming. More on that below.

One sign of the cultural interchanges involved, besides evidence for a highly mixed linguistic environment, Turkic names for Chinese cities, for example, including Qanbaliq, now Beijing 北京, are many physical artifacts, everything from art to maps to astronomical and medical treatises. Among the latter is the nearly 500 pages fragment, 15%, of what once was an encyclopedia of Arabic medicine, the *Huihui yaofang* 回回藥方 (*HHYF*), "Muslim Medicinal Recipes", or perhaps better, "Western Medicinal Recipes", so much is after all Greek. This is the title of the fragment at least.

In its present form the *HHYF* is a portion of a once 3200 page manuscript compendium compiled, based upon internal evidence, between 1398 and 1415, but almost entirely certainly based upon an earlier version from Mongol times. The full text when it existed was apparently intended to be a hospital notebook of a type well known from the Middle East.[11]

That the *HHYF* survives in only fragments is probably for a number of reasons, including the vicissitudes of transmission, but is also owing to the exotic nature of the medicine practiced within it. By Ming times there was still interest in Arabic medicine, but difficulty in obtaining so many rare medicinals. Also a factor was the more or less collapse of what was apparently a sophisticated medical infrastructure with the end of Mongol rule in China and East Asia, or at latest in early Ming times. The Yuan original of the text from which the fragments ultimately come, likely not called *huihui yaofang*, a name properly applied only to the Ming fragment, is from the late thirteenth century or the early 1300s. It is now completely lost. Fortunately, before it was completely lost, the Ming had the sense to republish it, or, more properly publish an edited version of the Yuan edition. It is some of this edition that survives, but showing

11 See the discussion of parallel texts in Lev (2013). See also Chipman (2010), and Lev and Amar (2008).

signs also of another extensive reworking later, probably in the sixteenth century when still fewer rare medicinals were available. The Ming by that time was no longer interested in the Indian Ocean world, the source of most of the medicinals, creating the need to offer many substitutions. Most of them are unworkable or even toxic. Clearly, by that time the medicine of the *HHYF* was no longer a viable medical system, in China at least.

Of the original 36 chapters of the early Ming (between 1398 and 1415) edition of the *HHYF*, three content chapters (*juan* 卷 12, 30, 34) survive, along with the Table of Contents for the second half of the compendium. This means that we have precise details about the contents of more than half the book (20 *juan*). In fact, our knowledge of what the *HHYF* once contained extends to more than 20 *juan*. This is because the surviving *juan* contain references to sections now entirely lost. Some gaps can also be filled in through comparisons between the *HHYF* and other sources for the Arabic medicine of the time. These include, as we shall see, probably Ilqanate sources, from Mongol Iran, although there may be a strong Indian influence at work as well. A full identification of sources awaits further research.

Each of the three content chapters is organised around one or more disease categories, the largest being the detailed discussion of "wind" ailments, a Chinese category in focus, but easily applicable to the Arabic world,[12] with various subcategories. It occupies all of *juan* 12, and then there are "various symptoms", a hodge-podge which occupy all of *juan* 30. By contrast, *Juan* 34 is comprised of shorter discussions of wounds from metal objects, broken bones, including a highly interesting section on head wounds and skull fractures, cauterisation, scalds and burns, wounds from blows, and bites, even human. The approach in these three chapters is overwhelmingly Arabic and supporting the discussions presented are numerous quotations from the various Arabic medical authorities. These include, and this is unique for East Asia, Zhalinuxi 扎里奴思 (i.e. Galen [os]),[13] but also, among others, Paul of Aegina, Hippocra-

12 But this is also an Ayurvedic category, on that more below.

13 Galen (second century AD) was from Pergamon in Asia Minor, once a major center of Hellenistic culture. He came from a solidly upper class family with a distinguished lineage stretching back several generations. Although other options were open to him, Galen became a medical theoretician, anatomist and physiologist and, last but not least, a practicing physician. His extensive travels eventually brought him to Rome where he had a distinguished career. Galen's importance above all lay in his interest in codifying and summarizing Hippocratic medicine, a huge tradition by his time. He was also very modern in his close interest in experimentation based upon daily dissections. Of his works, nearly 100 survive in Greek where they constitute approximately 10 percent of all Greek literat-

tes,[14] Rufus of Ephesus,[15] Aristotle, Alexander the Great, and, of course, the great purely Arabic authorities such as Ibn Sīnā.[16] Because of its nature, but unusual in China, the *HHYF* includes Arabic Script entries referring to the names of medicinals, recipes, and individuals cited, and even technical terms. These are also given in Chinese transcription in some cases showing the influence of Persian syntax, and perhaps an Uighur pronunciation both of Persian and of Chinese words. The transcriptions alone provide rich evidence of a mixed cultural environment where Chinese was just one of several languages in use.[17]

From the Table of Contents we know that, in good Arabic tradition, the *HHYF* once included a full discussion of *materia medica*, preceded with a general overview of the various types of medicinals used.[18] From this Table of Contents we also know the names of hundreds of simples and formularies not mentioned elsewhere in the text. Some are quite famous compounds

ure surviving from before 350 CE. Many more, lost in Greek, survive in Syriac and Arabic, or, in a few cases, in Latin. His works subsequently became the basis for medical theory and education not only in the Arabic world, but also in the European, including through those works passed back to Europe, via Arabic translations rendered into Medieval Latin. Galen was by no means the most brilliant physician of his time. but was one of the most wide-ranging, one reason for his importance. In any case, it was Galen who established the orthodoxy of the Hippocratic traditions of Greek medicine, and of his own views of them, almost to the exclusion of other schools. See Nutton (2007).

14 The Hippocratic corpus that Galen interpreted is associated with an historical physician, Hippocrates of Cos (460–375 BCE ca.) but none of the 60 writings in the corpus may go directly back to Hippocrates. Despite such uncertainty, the works in the corpus are important because they represent the best that the Greek world of the time had to offer to medicine. This includes the famous *On the Sacred Disease*, a rationalist attack on superstitions associated with epilepsy, and the first full discussion of the disease in any language. Despite their importance, and the free availability of Galen's own Hippocratic works in Arabic, including commentaries on unavailable works, the Hippocratic corpus is represented very unevenly in Arabic translation. Some of the most important works in the tradition are missing. See Ullmann (1978).

15 Rufus of Ephesus, 100 CE ca., was a major medical writer including about psychological topics but little survives in his native Greek. The situation is different in Arabic. Citations such as those in the *HHYF* thus usually preserve material that is otherwise lost, increasing its importance not just within the context of Arabic medicine. See Nutton (2004), 208–211, and *passim*.

16 See the complete list and discussion in Appendix 1.

17 This point has been developed in Matsui 松井 (n. d). See also Endo 遠藤 (1997). For an introduction to the multilingual character of the Mongol World see also Allsen (2001).

18 The medicinals, in the context of a complete translation of the *HHYF*, are analysed in full detail in Buell and Anderson (2021), also treated in more detail is the technical vocabulary.

and may be reconstructed using other sources. The Ming version of this text, the only one now surviving, or rather its fragments, exist in two manuscripts, one located in Beijing 北京 containing 4 *juan* out of what once were 36, as described above, and another even more fragmentary, in a regional library (not seen). The Beijing manuscript has now been published in two editions, one by S.Y. Kong (江潤祥)[19] and one by Song Xian 宋峴.[20] The latter edition includes an extensive Chinese-Arabic glossary and textual explanations in some depth.

Quotations from numerous Arabic Medicine authorities support material presented, including recipes. There are many theoretical discussions supporting textual material. They include direct quotations of various Arabic medical authorities in Chinese translations. Such quotations are by no means confined to Zhalinuxi, but other Greeks and authorities are not always named, and there may be more Ibn Sīnā (980–1037)[21] than meets the eye (Ibn Sīnā also apparently exists among unnamed authorities in the *Yinshan zhengyao*, the Yuan Dynasty imperial dietary, too). In any case such references are of potentially even greater

19 Kong 江 (1996).

20 Song Xian 宋 (2000).

21 Abū ʿAlī al-Ḥusayn b. ʿAbd Allāh b. Sīnā (980–1037ca.), usually known just as Ibn Sīnā or, in the West as Avicenna, was one of Arabic Medicine's greatest theoreticians. He came from one of the outlying areas of the Muslim World. By today's definitions he was born in Uzbekistan, of Tajik parents, Persian-speaking, in a place located near modern Bukhārā. Although an active official in the Sāmānid administration of his time for much of his life, he managed to find time to write numerous philosophical and medical works, making him one of the most important thinkers of his time.

Ibn Sīnā's most important medical work was the encyclopedic *Al-Qānūn fī al-ṭibb* "Canon of Medicine." It was comprised of five books. The first focused on medical generalities and theory. The second is a *materia medica* and herbal pharmacology. The third concentrates of pathology arranged by bodily systems. Book four dealt with fevers and a variety of other miscellaneous medical conditions, and the fifth with the pharmacopoeia of the entire collection. See Guichon in Waines (2011). What resulted was the most important medical book in East and West, although Ibn Sīnā's work was not always regarded as primary in the East. A modern translation, in five volumes, is Bakhtiar, et al. (1999–2014).

Ibn Sīnā's work is based in Hippocratic medicine as interpreted by Galen, many more of Galen's works than are now available in Greek. Much also reflects the authors of the Greek herbal tradition, principally the ever popular Dioscorides. Dioscorides was born in the first century BC in Anazarbe in Cilicia. He was known not only for his definite listings of *materia medica* (listings modified in the Arabic world because of differences in available species) but also because of his association of *materia medica* with a rich traditional of botanical drawings. The Arabs added their own versions and expanded the illustrated corpus. Dioscorides became a major foundation for the rich botanical literature of the Arabic world. See Dubler in Waines (2011). On Dioscorides see Sadek (1983).

importance given the upsurge of interest in that synthesiser and philosopher under the Ilqanate (Sara Nur Yildiz, personal communication to the author, September 24, 2011). Note that references to the Greek authorities in particular often include chunks of text, some of it substantial. Paul of Aegina is particularly well represented in this regard as in other Arabic medical literature.[22]

Although the Table of Contents for the second half of the book might seem to be the least valuable part of the surviving material, this section provides us with very important information: the overall structure of the text as a whole, or at somewhat more than half of it, with *juan* 12 and the cross-references to lost parts. From this Table of Contents we know, for example, not only that the HHYF once included a full discussion of *materia medica*, but one preceded with a general discussion of the various types of medicinals used in its Arabic medicine. Thanks to the partial Table of Contents, we also know the names of simples and hundreds of formulas not mentioned elsewhere in the text. Some are of the latter are quite famous compounds found in other sources, sometimes in more complete form. We can also tell a great deal from the order of the presentation of subtopics in the theoretical discussion as indicated by subentries cited in the Table of Contents, but not surviving.

In its present form, and we have no reason to assume that the lost sections of the HHYF were all that different, and looked at as a whole, the text offers three distinct types of material. The largest component, perhaps making up sixty per cent of the content chapters, are medical recipes, some quite elaborate, and many quite famous. These recipes are intended to respond to one or more specific conditions, and are often associated with a detailed symptomology. Next in quantity, perhaps making up thirty per cent of the whole, are theoretical discussions, sometimes of types of injuries or illnesses in a larger context, and including in passing some recipes or procedures. Included are extended citations in some cases from the HHYF's medical authorities, including Galen, but also Hippocrates and a variety of Arabic commentators. Although not a primary thrust in the surviving sections of the HHYF, materials in this area include a few detailed discussions of specific medicinals. An example on castoreum is given below.

22 Paul of Aegina was a seventh century (625–690 ca.) Alexandrian physician and medical writer most known for his great encyclopedia *Pragmateia*, "Pertaining to Practical Things", which survives (along with other works) and made Paul an authority for much of the Byzantine period. On the Arabic tradition of Paul, see now Pormann (2004). Also Gutas (1998).

Finally, making up about 10 percent of the whole are practical discussions of first aid for a variety of injuries, e.g. burns, breaks, even skull fractures, and bites. As above, this section includes frequent references to specific recipes and preparations, not all of them internal. The information on splinting is particularly full, as is the discussion of various ways to set broken or dislocated bones associated with various parts of the body.

The following examples are typical of the material in the text. The first is a description of a medicinal in Arabic terms, in this case castoreum with full details including, very Arabic, how to distinguish the fake from the real, and how to treat poisoning that can result from using bad castoreum. Note the mix of Persian and Arabic terminology, even to the use of a technical expression, *sarsān*.[23] There are hundreds and hundreds of such expressions in the text as a rule used correctly and in a proper context:

> For an ingredient take *gundbidastar* [castoreum], another name is *mīyān-e khizā* ["middle of (beavers) testicle."]. This is castoreum. Rub this material on the body. It can cause body heat to dissipate and disperse its wind. All castoreum has a pair of attached skins. This is the real thing. That with only a single attached skin is mostly fake. The fake uses *Jāwashīr* [resin of *Opoponax chironium*] ([Subtext] *Jāwashīr*) and *ṣamgh ʿarabī* [gum arabic, including from *Acacia sperocarpa*]
>
> ([Subtext] This is a resin found in a plum tree in the Arabī land). [To make it] take a little castoreum and grind up completely. Combine with blood and store inside of a bladder. When sun dried then it holds the false castoreum. Also, castoreum, in terms of is original nature, is third to fourth rank in being heating, is class two in producing drying. Also, when castoreum has a dark color approaching black, it is then a poisonous substance. If people take it, they must have damage from this. However, although it does not do damage, it will produce *sarsān* [brain membrane fever] head swelling symptoms ([Subtext] Also symptoms of mental confusion). One needs to counteract the poison. Take juice squeezed from an orange, or vinegar made from grapes, or donkey's milk and consume it.

---

23 The Arabic and Persian names are usually used, in Chinese transliteration, with the Arabic or Persian script often supplied as an annotation, that is, as a subtext in smaller characters. Script entries in turn are transliterated into Chinese. The Ming edition we have displays good to excellent calligraphy in these languages—there must have been actual Arabic and/or Persian scribes at court. The transliterations are usually clear enough if one allows for linguistic changes since Mongol times, and makes provision, in many cases, for a filtering through a Turkic language.

> All can counteract [the poison]. If one uses other medicinals, do so along with these substances. If one lacks these substances it is also possible to use sweetflag rhizome or black pepper, half the quantity, to replace them (*HHYF* 12, 142–143).

All medicinals used in Arabic medicine are treated similarly to castoreum when they are analyzed, some in even more detail. The result is that medicinals are well-known and described within the tradition, including all medical properties. The book tradition of Arabic medicine, which draws on all of this, is thus replete with carefully balanced recipes designed to respond to specific conditions, and expressed in terms of imbalances within the traditional physiological system.

In the following two examples, the type or derivation of each recipe is given, followed by a consideration of the specific application. There then follows a listing of ingredients, some of them quite rare, and instructions as to how to prepare the dose and, in many cases, how to employ it. By *Qi* understand Greek-Arabic *pneuma*, but the usage here is also perfectly suited to Chinese conceptions of *qi* 氣. In the texts, a *qian* 錢 is a Chinese ounce, or 3.12 g today, and is one-tenth of a *liang* 兩. A *fen* 分 is one-tenth of a *qian*. In a text in Arabic the measures would, of course, be traditional Arabic measures for medicines, but here adapted to Chinese usage. Note the association of the second recipe with Galen himself. As in the previous example, the subtexts include Arabic script entries:

1. *Maʿjūn* ([Subtext] This is an ointment medicine) [Ar.] *Dhū al-Misk* [having musk, aromatic] ([Subtext] *Dhū al-Misk*)
   It clears out wind and accords *qi*, it helps move *qi* force and pacifies the heart. Use:
   Cynanchum
   Idkhir [lemon grass, *Andropogon schoenanthus*] ([Subtext] *Idkhir*)
   Fennel
   Chinese spikenard
   Common rue ([Subtext] Each three *qian*)
   Tsaoko cardamom
   Cloves ([Subtext] Each two *qian*)
   Zurunbād [wild ginger, root of *Curcuma zerumbet*] ([Subtext] *Zurunbād*)
   Durūnj [leopard's bane, *Doronicum pardalianches*] ([*Subtext*] *Durūnj*)
   Amber
   Pearl ([Subtext] Each two *qian*)
   Musk ([Subtext] One *qian*)

Piannao 片腦 [*Dryobalanops aromatica*] ([Subtext] half a *qian*)
Grind up the ingredients together into a fine powder. Refine honey and combine into an ointment.
Each time take one *qian*. Send down with honey water. Together there are 13 flavors (*HHYF*, 12, 12: 147).[24]

2. Al-Jālīnūs [Galen] ([Subtext] This is an ancient Muslim medical man) proven recipe. It can treat tendon artery swelling epidemics such as [Ar.] *faliqamūniyā* [phlegmonia] swelling ([Subtext] *faliqamūniyā*). If used it will be aided.
Qalqadīs [red oxide of iron, iron pill] ([Subtext] This is dried yellow potash alum. One *qian*, one and a half *fen*.)
"Golden thread soda" ([Subtext] Nine *qian*, four and a half *fen*)
Copper powder ([Subtext] Two *liang* two and a half *qian*)
Frankincense skin ([Subtext] One and a half *liang*)
Bārzad [galbanum, resin of *Gerula galbaniflua*] ([Subtext] *bārzad*. One *liang*)
Yellow wax ([Subtext] Seven *liang*)
Zayt [olive] oil ([subtext] This is oil produced from the *zaytūn* tree [olive tree] tree of the Shām [Syria] land. Nine *liang*)
Grape liquor vinegar ([Subtext] 45 *liang*)

First take the dry medicines and grind for ten days. Afterwards take the powdering medicines and after powdering, combine. Propagate to the entire system where there is a wound. Do it twice a day, or three times. When using the medicine also add warm or hot *zayt* [olive] oil on top. After that take a fur cloth and dip the vinegar with the *zayt* [olive]. Warm-hot apply to around the place that is wounded. Moreover, one must tabu cold and chill. The reason is that that which the tendon main arteries fear is only chill together with stiff things (*HYYF*, 34: 435–436)

Of the two recipes, the first is purely for internal use, the second for external use, where there is a wound, but both are compounded in accordance with exactly the same theoretical system. But this is not to say that the tradition contains no procedures for such things as broken bones, in the case of the

24 All citations of *HHYF* are from the edition of Kong 江潤祥, et al. See Kong (1996). All translations are by the author and are strictly copyright.

HHYF, even for skull and brain injuries, conditions harder to make a direct part of the humoral system, but still part of it by implication. This is true, for example, in terms of recipes to follow up a bone splinting to overcome the shock of a bone break, and allow the rebuilding of bone and muscle. Also a key part of Arabic medicine in the medical books is a well-developed dietary tradition, also coded humorally. No clear distinction was made between food and medicine. Most of what we think of as spices were, for example, both, and treatment could call for what seems to us to be a purely dietary treatment.

When such material is looked at in detail, a number of observations are called for. One is how rooted in Arabic medicine most of the material in the *HHYF* is, although not exclusively so, and there has been a clear effort to assimilate the Arabic material in the *HHYF* to Chinese, and most likely other medicines. In the stand-alone recipes, for example, the *materia medica* is overwhelmingly from the Islamic World, universally using the appropriate Arabic or Persian names for the substances. The overall presentation of recipes, and their titles, is also Arabic and it is intended to respond to specific conditions recognised in Arabic medicine, some even named in Arabic, and well known elsewhere. This impression is strengthened when one looks at other sections of the *HHYF*. The theoretical discussions and like material, for example, are almost entirely formulated in terms of the traditions of Arabic medicine, and when Greek authorities are named these are probably not from any original text in particular, but from the rich Arabic translation literature from Greek medicine in general.

Only the first aid material seems less Arabic but this material too cites Arabic recipes and preparations as well as authorities. Also Arabic are many of the specific considerations of the text. The cauterisation section in *juan* 34, for example, is entirely unchanged from an orthodox Arabic Medicine text. The Chinese do burn in this manner under certain circumstances, but the methods here are entirely Arabic. Discussions of individual medicinals also appear to come straight from the textbooks of Arabic medicine containing similar material. At first glance, the only real Chinese touches in any of this material are a number of attempts to explain very foreign medicinals and the like to a Chinese audience. That so many of these are completely off, just plain wrong, probably tells us something very important about the dating of the Ming edition of the *HHYF*, namely that the present primary copy probably represents a revision of the early Ming edition and came from a period when the free flows of medicinals by land and sea had, at least temporarily dried up, or been reduced, that is, most likely from the period of the Ming overseas trade ban after the Zheng He 鄭和 voyages.

But as Arabic as the material may appear at first glance, appearances deceive. One indication that not all is orthodox Arabic medicine[25] lies in the area of the area of the humors referred to in the text. As I have shown elsewhere,[26] instead of the expected four humors of Arabic medicine, e.g. blood, phlegm, yellow bile and black bile, the system employed in the *HHYF* instead appears to involve only three, air (*qi*, more often than not, not used in the Chinese sense), bile (in this case yellow bile, yellow liquid), and phlegm. Blood is mentioned but is clearly not a humor. There is to be sure a "black blood", but this cannot be black bile, never associated with blood in the Arabic system, and this "black blood" is not a humor.

Whence this strange system? Perhaps the origin is not all that distant since Ayurvedic and Tibetan medicine both have a system of three such humors, and not the four of the Arabic world.[27] Making this all the more likely is known Tibetan influence in the parallel *Yinshan zhengyao*, the imperial Yuan dietary manual of 1330, which also incorporates substantial influence from Arabic medicine.[28] There is also the known presence of Tibetan doctors and other medical practitioners in high positions at the Mongol court which almost certainly produced the *HHYF*,[29] and the known existence of a parallel tradition of "Persian" medicine in Tibet, represented at the Mongol court in China.[30] This included among other Tibetan doctors present in China, some of those very Tibetan doctors involved with the traditions that ultimately became the *YSZY*. Interestingly, one of the surviving texts of the school is a specialised manual that deals with head trauma, closely paralleling what is described in the *HHYF*.[31] Also possible is more direct Indian influence. The use of "wind" in the text can and most likely should be interpreted in Chinese terms but "wind" plays a major role in Ayurvedic medicine.

Not orthodox Arabic medicine are also references to *qi* in what is clearly a Chinese sense, such categories as 7 apertures of the body, a completely Chinese concept, and frequent mention of *jing* 經, sometimes major blood vessels, but also, apparently, and this is ambiguous, the channels through which *qi* flows,

25 On the general subject of Arabic medicine see Pormann and Savage-Smith (2007), Dols (1984) and Ullmann (1978).
26 Buell (2010).
27 See now Speziale (2018), Speziale (2014a) and Speziale (2014b).
28 Buell and Anderson (2010).
29 Buell (2010).
30 On the various foreign traditions present in the Tibetan medicine in China cfr., as an introduction, Martin (2010). I offer particular thanks to Olaf Czaja for sharing the results of his unpublished research in this area with me.
31 Buell (2010).

although this certainly is not the basic meaning in the *HHYF*. Many of the disease categories, although most are not without their equivalents in Arabic medicine, are purely Chinese, e.g. *feng* 風, "wind", when used in an entirely Chinese sense to describe strokes and similar conditions. The *HHYF* is thus a complicated text with several medical traditions present and considerable evidence for assimilation, and not just to one alternative tradition.

Given such realities, what are the origins of the *HHYF*'s materials? Clearly, medical texts focusing on Arabic medicine were important since Arabic medicine is the pervading influence in the text as a whole, despite an effort to adapt the material to Chinese terminology. There is also some evidence for assimilation to a medical environment in which more than one type of doctor was practicing. Thanks to the Uighurs, who were a strong secondary presence as the *HHYF* was compiled, this environment may already have represented a highly assimilated tradition of medicine when its Arabic side first came to China.

Editor S.Y. Kong and his team have emphasised the relationship of the text to the great *Qānūn fī al-ṭibb* of Ibn Sīnā. This was a standard medical encyclopedia in the Islamic world, at least at the time that the *HHYF* was compiled in China, but this relationship is probably a distant one at best. The *HHYF* certainly reflects a renewed interest in the *Qānūn fī al-ṭibb* but the *Qānūn* is not even its most important source and it is a mistake to regard the *HHYF* as little more than a Chinese adaptation of the *Qānūn*.

Some sources may be surmised. One is probably Al-Bīrūnī (973–1050) but at one or two removes, but this remains to be investigated in depth.[32] Shams al-Din al-Samarqandī (thirteenth century), who wrote in temporal and physical proximity to the *HHYF*'s original, seems surprisingly unrelated. But one possible source appears to have been Sayyid Ismāʿīl Jurjānī's *Dhakhīra-i-Khwārazmshāhī*, "Thesaurus of the Khwārzamshāhs", a monumental twelfth century Persian work very important in Central Asia at the time. It was apparently written there.[33] If the *HHYF* did not rely on the *Dhakhīra-i-Khwārazmshāhī* directly, then it likely did on one or more of its own probably largely lost sources.[34]

32 Cfr. Said (1973).

33 Elgood (1951), 216–218.

34 Conspicuously missing as an authority among those mentioned (cfr. Appendix 1) is ʿAlī ibn-al-ʿAbbās al-Majūsī (fl. late tenth century), author of *Al-Kitāb al-Malakī*, "Royal Book" a theoretical work that was one of the largest of its kind. In it, Al-Majūsī generalised about illness and human physiology at four levels: the Greek elements and their extensions, generalised to embrace the entire human system, the temperaments, the four humors, and the faculties. Also important are the *pneumata*, the link between the strictly material and spiritual side of man. At their simplest level, the elements are the traditional fire, air, water and earth, the primary or homogenous elements. To this are opposed vari-

One thing is certain, and this ties in well with the possibility of Jurjānī as a major source, the overwhelming majority of the sources for the *HHYF* appear to have been written in the Persian language rather than in Arabic. Persian is clearly the working language of the text. Arabic script entries are, more often than not, Persian grammatically. The characters chosen for Chinese translation also indicate Persian readings of the words. But not all sections were from Persian sources. The listings of *materia medica*, do appear often Arabic, but such sound shifts as use of "j" sound for ق may indicate passage through Turkic hands (the sound shift is probably a Turkic palatalisation, and the same change is found in the *YSZY*, where there is pervasive Turkic influence). Apparently Turkic words and forms also exist in the text. Further research is needed.

---

ous secondary and even tertiary elements which can be further grouped into terms of immediate or a more distant connection with physiology. In practice, in terms of this system, the homoeomeral organs of the body, that is, the basic organs from which all other organs of the body are composed, are also elements, as are the humors themselves which make up the homoeomeral organs. The system is further tweaked by Al-Malak to express nearly also aspects of basic physiology. The elements of the body are also formulated in terms of the hot, cold, wet and dry system, a basic categorisation of humoral medicine.

At the next level, the temperaments reflect the precise mix of hot, cold, wet and dry. In humans, there are nine basic temperaments, mixtures of hot, cold, wet and dry, one in balance and eight out of balance. The latter are comprised of four simple temperaments and four composite temperaments. But there are many more variations due to place, habits, sex, food, and so forth.

The four humors are secondary elements, that is, they exist above the level of fire, air, water and earth, the primary elements, but are still closely connected with them. In Arabic medicine the humors, blood, phlegm, yellow bile and black bile, are each associated with an element and have properties in terms of the hot, cold, wet, dry system. Phlegm is cold and wet, for example, with some internal variations, and corresponds to the element water. Although not all can change, some humors can change into another under the right conditions, e.g. phlegm can become blood when worked on by an "innate heat." The real importance of the humors for the system is that the dominance or absence of a given humor results in illness. And when the dominance or absence becomes extreme, death can result.

The faculties are three fold: 1) natural faculties involved in conception, physical growth and the process of nourishment; 2) animal faculties most closely associated with the process of life, in the beating of the heart for example; and 3) psychic faculties associated with such things as reason. Each set is associated with a particular part of the body, the psychic faculties with the brain, for example. And each set of faculties is associated with a kind of *pneuma* with the relationships involved strongly reminding of the complex system of *qi* 氣, a close equivalent of the Greek (plural) *pneumata* important in Chinese medicine but with nothing apparently to do with Greek ideas.

All of this has been noticed by S.Y. Kong's team. They strongly suggest that the *HHYF* was compiled primarily by Turkic-speakers, probably Uighurs. This might be expected, given the known influence of Turkic-speakers at the Mongol court. Thus the *HHYF* may be one more expression of a Turkistani tradition of Arabic medicine that is otherwise largely unknown and unstudied but which must have been major and highly creative in Mongol times. There likewise appears to be a Syrian-Nestorian presence in the text, a fact that may be connected with one of the probable go-betweens for material in the text, as indicated below. Neither of these influences is unexpected given what we know about the cultural flows of Mongol times in particular.

Also clear is that the *HHYF*, although primarily documenting the Islamic medicine of the time and reflecting contemporary Persian and probably Turkic usage, it is also an important document for Chinese culture as well. This is not just in terms of the assimilation of some of the medicine and ideas involved. The text is, for example, written in a type of colloquial Chinese known from other Yuan Dynasty-era documents. There is considerable internal evidence that the Chinese language of the text is a specialised, technical language with considerable apparent historical depth. In any case, the text is by no means a literal rendering of its Persian and Arabic sources but reveals considerable effort, not only to express and translate Islamic ideas, but also to assimilate them to the concepts of the Chinese medicine of the time, including the key concept of *qi*, the "life force" making traditional Chinese physiology work, although the *qi* of the *HHYF* is often a "humor", "air", as well. This usage seems to reflect what I see as a major Tibetan (and thereby Indian) influence. But often the text's *qi* is just Chinese *qi* and nothing else, showing how much the *HHYF* has been assimilated to Chinese tradition.

The Mongol period, along with late Song 宋 (Southern Song, 1125–1279) and Jin (1125–1234) times, was a high point in the development of traditional Chinese medical theory. A close study of the Chinese language of the *HHYF* could thus not only expand our understanding of the text considerably, but also improve comprehension of the particular Chinese medicine of the period.

Such then are the obvious influences, and much more will be identified as research continues. These almost certainly include a substantial south and Southeast Asian element in the text through the specific *materia medica* called for by the *HHYF*. These are often quite different in mix than those in use in contemporary Cairo, for example, as studied by Chipman,[35] but not exclusively so. Elsewhere I postulate that this is connected with the great interest of Qubilai-

35 Chipman (2010).

qan and his successors in the Indian Ocean in an attempt to link with Mongol Iran, Mongol China's most important ally in the struggles of the wider Mongolian world.[36]

We may easily postulate how trade alone brought many of the *HHYF*'s medicinals to China. This given the interests of traditional China in a far-reaching trade in exotics, which became more exotic the more distant their origin. Nonetheless, trade alone cannot account for other elements of the *HHYF*. These include, most notably, theoretical ideas. Even if there is substantial Chinese or even Tibetan influence, the substratum is Arabic. How did such material, such specific material in many cases, make it way to China? Some no doubt had long come via the Uighurs, but not all is Turkic, apparently just a small part of the whole, even though Turkic speakers may have been among the translators and adaptors of Arabic or more likely Persian material.

In terms of the rest of the material, one individual must have been particularly significant. This is the Syrian ʿĪsā (Aixie 愛薛), probably speaking a variant of Syriac. ʿĪsā was an important figure in Qubilai's China where he held high ministerial posts, and in Mongol Iran, where he went as an ambassador (and returned).[37] And there is more, ʿĪsā is not only is known to have been associated with the Arabic medicine institutions of Mongol China, including a public charity clinic, he and his family that is, but to have been a doctor himself, and to have actually founded these very institutions already existent in the thirteenth century. They in fact belonged to Isa's family before they became national institutions. Thus, as much as anyone, ʿĪsā introduced Arabic medicine to the Mongols, in an activism dating back to Mongol imperial times, and, most important, was a known go-between between Mongol China and Mongol Iran and back again. Is it so farfetched to assume that he came back to China with a satchel full of books, including editions and new interpretations of *Ibn Sīnā*'s important writings?

There is another smoking gun of long-distance association, and that is the known interest of the Ilqāns, principally the translation bureau of the great minister Rashīd al-Dīn (1247–1318), in Chinese medicine. This is reflected, among other places in the encyclopedia *Tansūq-nāma-i Ilkhānī dar funūn-i ʿulūm-i Khitāy*, "Treasure book of the Ilqans on the Sciences and Learning of China", which translates and interprets a great deal of information regarding Chinese medicine, including an entire Chinese treatise on pulsing, as well as

36 Buell (2012). Cfr. also Buell (2009). The Indian Ocean was, of course, a major conduit, for medicinal trade as well. Cfr. Buell (2014).

37 Weng, 1938. Cfr. also, in passing, the discussion in Allsen, 2001, and also Buell (2007).

loose information from such sources as the *Yellow Emperor's Inner Classic*.[38] What is especially interesting are parallels in content and style between the *Tansūq-nāma* and the HHYF suggesting that the two texts were prepared in much the same way, with reading and interpretation from a written source or, sources. The one set of sources was presumably in Chinese, the other probably in Persian primarily. In the case of the HHYF, there is also the clear suggestion that the readers of the source texts were Uighurs, something that remains to be proven for Mongol Iran, but which we strongly suspect. We do not know for sure but the conservatism of some of the transliterated forms does suggest contemporary Uighur usage of the era.[39]

The HHYF thus did not exist in a vacuum, and appears to have been part of an active exchange, one whose full scope we still do not understand; so much is lost. Also a problem is that the material in question has been little researched, including the *Tansūq-nāma* itself, although this may be changing. There may even be more Chinese texts surviving in unique copies, or very small editions in Chinese libraries and specialised collections, whose existence is of yet barely suspected. Such texts certainly once existed, possibly included among books copied out into the early Ming encyclopedia *Yongle dadian* 永樂大典, if its Arabic medicine section is not the full version of the HHYF.

The HHYF is our principal surviving source for the Arabic medicine transfers of the Mongol period, but it is not our only source. Another text heavily influenced by Arabic medicine, although the influences are not always so directly obvious, is the imperial dietary manual, YSZY. It was presented to the Mongol court in China in 1330 by an author who, if not from a Turkic-speaking minority, was heavily influenced by Turkic culture and draws upon various Arab material along with Chinese classics of Taoist lore and dietary medicine. The text, for example, includes various Arabic *sharba*[40] intended for medicinal purposes,

---

38 Cfr., as an introduction, Klein-Franke and Zhu (1996), and Klein-Franke and M. Zhu (1998). Like the HHYF, the *Tansūq-nāma* as it survives today is only a fragment of what once was a much larger text and some of the material seems even to have been lost shortly after the work was compiled since someone early supplemented lost sections with material from other sources. Cfr. Klein-Franke and Zhu, 1996: 399–400. Note that, as in the HHYF, much of the material in the *Tansūq-nāma* has been adapted with Arabic humoral medicine in mind, even to the extent of distorting the Chinese content of the text in some cases, e.g. the so-called Rashīd Al-Dīn introduction, for example. Cfr. Klein-Franke and Zhu (1996): 400–401. Cfr. also Buell (2007).

39 Matsui (n.d.).

40 One is even called that, made with red currants. Cfr. Buell, Anderson, and Perry (2010), 374–375. In fact the entire section from which this recipe comes shows overwhelming Arabic influence, even if not openly labeled as such.

as well as for their use as sweet drinks. In fact almost all of the recipes calling for substantial amounts of sugar show Arabic influence. The text also contains interesting references to saffron as cheering based in Arabic humoral theory.[41] Likewise, some of the *YSZY*'s many prohibitions closely parallel information in the dietary medicine of the Arabic cookbooks of the period.[42] Even more striking, is a whole section of the text that appears straight out of Ibn Sīnā, namely the long discussion on food and other avoidances for a wet nurse.[43] Such material exists in Ibn Sīnā, but was not part of the Chinese tradition at all, at least prior to the *YSZY*.

One further point on *HHYF* influences. Greek medicine is extraordinarily important in it, both in a derived and an original form, e.g. the many direct quotations, sometimes big chunks of text, from Paul of Aegina and others. Possibly we will be able to collect and trace more references still at some future date.

In any case, Arabic medicine as it is in the *HHYF* is by its very origin associated with the littoral of the Mediterranean Sea, and Fernand Braudel long ago emphasised[44] the essential unity of the societies and cultures located along it. This is in large part due to the common experiences of all in such essentials as agriculture, food, trade, art and architecture. From the Mediterranean, in turn, paths lead off into associated realms not so directly connected. These include interior Europe, what is now Russia, via the Black Sea, inner Syria, Iraq, Nubia, and interior Africa and so on to even more distant places. The so-called Silk Roads, for example, long linked the Black Sea littoral and Iran with Central and East Asia. From Egypt the Red Sea leads to the Indian Ocean, increasingly important after the thirteenth century, while the world of India abuts Iran directly, and was accessible by sea thanks to the monsoons. Among the cultural goods travelling along the Mediterranean Littoral, and via various routes linking the nearer and farther worlds associated with the Mediterranean, has been medicine.

At first multiple medical traditions were present. Each of the major cultural areas located along the Mediterranean had its own traditions of treatment, its own herbs and other *materia medica*, and even magical practices, although much in any given area was related to what was found in the others. In ancient Egypt much more than a local medicine was involved, as witnessed by the consistent and theoretically well-grounded practices found in the papyri. Farther

41 Buell, Anderson and Perry (2010), 551.
42 Cfr. for example material in Nasrallah (2007).
43 Buell, Anderson and Perry (2010), 265–267.
44 Braudel (1998).

afield, in Mesopotamia, several major medical traditions came into being at a comparative early stage of regional development. Mesopotamia traditions also influenced those of Syria, the Hittites and Iran.

Nonetheless, despite all the divergence and regionalism, a good deal of which still exists locally, the Greeks were able, in the end, to put together a system that, much altered, still dominates. It was not without foreign influence, and the ability of Greek medicine to absorb much coming from outside was one of its strengths.

Greek medicine had two roots. One lay in Greek relationships with the gods and the personified forces of nature. This included Apollo, for example, who cast a plague upon the Greek armies besieging Troy. The other lay in a rationalistic, largely non-religious interpretation of disease and illness based upon a view of the universe interpreted in strictly materialistic terms. On one side was the world of the gods, the other of the *iatrós*, the physician. This is a word already found in Mycenaean texts but strongly represented in Homer with the intensely practical approach to wounds and injuries in the *Iliad*. The world of the *iatrós* was also important for the thinkers and philosophers known as the pre-Socratics, and is associated with the complex of texts that later known as the Hippocratic corpus. This corpus was the work of many authors, and reflected many influences and points of view. Its texts were written over a considerable period of time. It represents the first fully rational treatment of medicine in consistent theoretical terms, although true consistency was only arrived at over time.[45]

Important for the history of Arabic medicine was the fact that earlier Hippocratic ideas were synthesised and reworked by the great figure of the whole tradition of Greek medicine, Galen (129–216 ca.). It is Galen's synthesis which has come down to use as the mainstream of Greek medicine. And it is this synthesis which was taken over by the Arabs as the foundation for their own medical literature, although this is not to say that Galen represented the only tradition present during his time. Medical research and literature quite antagonistic to Galen, although often fragmentary, has reached us. Some even reached the Arabs through anthologies of medical literature that were as typical of the Graeco-Roman, or Byzantine worlds, as they became for the Middle East.

Greek medicine is, without question, a pervasive influence in Arabic Medicine in general, and also in the *HHYF*. To be sure, it is only one of the influences present, but a very major one. In the end it is a backdrop that helps find our way in the individual texts of our fragments, from known to less known.

45 Nutton (2004).

Nonetheless, even with substantial Greek medical influence to guide us, the *HHYF* remains a difficult set of documents. Perhaps the biggest problem in dealing with the *HHYF* is that most of the illness terms are completely incomprehensible, at least to those not scholars in the field, or practitioners of today's variety of Arabic medicine. Although much remains purely in the form of technical expressions occurring in their original forms, in Arabic or Persian, too many Near Eastern concepts and terms have also been translated into China's medical language, meaning that their meanings must be puzzled out from context. Yuan was, to be sure, a time of medical change and transition,[46] further complicating things, and there are questions concerning what the terms of the *HHYF* really mean, especially when Chinese ideas are used to translate Arabic and Persian terms that are of uncertain referent themselves.

Even the text's more or less Chinese medicine, such as it is, can prove difficult. Chinese Medicine has a very different basis than Arabic. As noted above, it is based on *qi*, which literally means "breath", and by extension is, like Arabic Medicine's *pneuma,* a vital essence or vital energy, in part what makes the whole medical system work. There the resemblances stop. In Chinese medicine, and to some extent in our text, conditions are blamed on such things as stagnant *qi*, evil *feng,* inadequacy or weakness of the organ fields, and other causes unclear to us. In modern Chinese medicine, these entities can be diagnosed on the basis of suites of symptoms that have little or nothing to do with modern biomedical reality. And when you mix it all with Middle Eastern concepts, we are in very deep water, over our heads.

Even the relatively straightforward medical terminology used in the section on wounds, breaks and dislocations is often incomprehensible today, although anatomy in this case allows us to gain somewhat of an idea about what is being talked about. Bones and muscles have a certain reality to them, whatever the terminological haze. And *HHYF* doctors knew what they were doing and were working in real terms. That helps too.

The vast majority of the remedies in the *HHYF* are inherited, meaning in some cases that specific biomedical applications, even when generally known, may no longer have relevance, at least in the specific systems of the text. Most remedies were already Greek and/or Near Eastern long before they became Arabic. Many go back to Dioscorides, and, in fact, most of the specifics and medical ideas of Dioscorides, along with, as mentioned above, a great deal of

46 Cfr. as an introduction to Chinese Medicine, and the period, Unschuld (1986) and Unschuld (1985).

other Greek tradition are found in the *HHYF*. A few are Indian. Very few are strictly Chinese even though there has also been an on-going attempt to adapt the text to Chinese traditions.

To conclude, the *HHYF* is a unique document, one that is Arabic Medicine on the surface, but in fact shows many other influences, not just that of mainstream Arabic Medicine. The Mongol Empire was truly a world empire, and incorporated every bit of knowledge it could find; doctors from the West, Tibet, and China were present at court, and probably Indian medical experts too. Globalisation is, as indicated above, not new; it was not even new when the Mongols brought it to great height. How broad it could all become is shown by the immigrant physician of the Tibetan court in the mid-eight century. He was a Greek from the Byzantine Empire who called himself, appropriately, Galen![47] Texts claiming to be from the real McCoy can be found in the *HHYF* and more.

Appendix 1: Authorities in the *HHYF* (by Order of Appearance)

Yaḥyā ibn Māsūya[48]
Qaysar [Caesar]
Jālīnūs [Galen]
Rūfas [of Ephesus]
Aegina [Paulos of]
Abū ʿAlī [Ḥusayn ibn Sinā]
Mathrīdīṭūs [Mithridatis of Pontus?]
Sābūr [Sābūr ibn Shali][49]
Fawlus [Paul of Aegina]
Arkāghānīs [Archigenos of Aparmea, first century AD]
Qubādh [Persian king]
Athānāsiyā[50]
Arisṭūmākhis [?]
Arisṭāṭālīs [Aristotle]

47 Garrett (2007).

48 This is, apparently, Yaḥyā ibn al-Biṭrīq, who flourished during the first part of the ninth century. Cfr. Levey (1973), 139. Song understands the Chinese *Laoadiya* 老阿的牙 as [Ar.] *lawghādhiyā*, the name of a medicinal, but not, as the text indicates, of a doctor. I take the *lao* 老 in the meaning “venerable.”

49 Died 869. He was a Persian Christian. Cfr. Elgood (1951), 92.

50 This is apparently Athenaeus of Attaleia, cfr. Nutton (2004), 202–203. He lived in the first century BCE.

Filighuryūs [Philagrius of Epirus, third-fourth century AD]
Kisrūi [Crosroes Anūshirwān]
Ṣaḥārbakht [?]
Firifūryūs [= Filighuryūs?]
Aḥmad Fārruq [?]
Sarāfiyūn [Ibn Sarābiyūn Da'ud?]
Tiyādarīṭūs [Name of king. A Theodorikos before Galen]
Yūnāni man [a Greek]
Falāsifa [A Philosopher]
Ḥunayn ibn Ishāq[51]
Iskandar [Alexander the Great]
Buqrāt [Hippokrates]
Usquf [?]
Dalilasha 荅里剌沙 [Darius?]
Salamūyah [ibn Banān]
Umḥammad ibn Zakariyā' [960–932]
Rūmī Fulūniyā Ṭarsūsī [?]
Qubādhu [al-]malikī [see Qubādh]
Filūniyā-i Rūmī [= Rūmī Fulūniyā Ṭarsūsi?]
Abū Yaḥyā al-Marwazzī [? = Yaḥyā ibn Māsūya]
Dalimaxi [?]
Barmakī [Buddhist priest, eighth-ninth centuries?]
Sāmānī [Man of the reign of *Sāmān*, a Zoroastrian minor king of Khurāsān]
Jāṯhalīq [Christian doctor or priest]

51 Ḥunayn b. Isḥāq (808–873) was the most noteworthy for his translations of the works of Galen. Ḥunayn was born in al-Ḥīra, son of a pharmacist. He grew up culturally if not religiously Nestorian, and knowing both Arabic, the language of his native town, and Syriac, a language in which a substantial Greek-based medical literature already existed. Ḥunayn apparently preferred to use the latter language, at least more of his translation started out in Syriac than Arabic. According to Ḥunayn it was not so much a matter of personal preference as due to the fact that a specialised medical terminology already existed in Syriac, but not yet in Arabic. Among his translations were some 100 devoted to the works of Galen but this not all of those works listed by as translations of Ḥunayn were actually by him, and available lists are incomplete. In addition to his translations, Ḥunayn has also left behind various philosophical works in many subject areas, all showing his great familiarity with Greek. Some of these works were later translated into Latin [Cfr. Strohmaier in Waines (2011) and Gutas (1998)].

Bibliography

Allsen, Thomas T., *Culture and Conquest in Mongol Eurasia* (Cambridge: Cambridge University Press, 2001).

Bakhtiar, Laleh, et al. (eds. and transl.), *the Canon of Medicine* (*Al-Qānūn fī al-ṭibb*) (Chicago: Great Books of the Islamic World, 1999–2014).

Braudel, Fernand, *Les Mémoires de la Méditerranée, Préhistoire et Antiquité* (Paris: Editions de Fallois, 1998).

Buell Paul D. and Francesca Fiaschetti, *Historical Dictionary of the Mongol World Empire*, second revised and expanded edition (Lanham, MD: Rowman and Littlefield, Publishers, 2018).

Buell, Paul D. "Qubilai and the Indian Ocean: A New Era?" in Salvatore Babones, and Christopher Chase-Dunn (eds)., *Handbook of World-Systems Analysis* (London and New York: Routledge, 2012), 42–43.

Buell, Paul D. And Eugene N. Anderson, *A Soup for the Qan: Chinese Dietary Medicine of the Mongol Era as Seen in Hu Sihui's Yinshan Zhengyao*, appendix by Charles Perry, second revised and expanded edition (Leiden and Boston: E.J. Brill, 2010).

Buell, Paul D. And Eugene N. Anderson, *Arabic Medicine in China: Tradition, Innovation and Change*, (Leiden and Boston: E.J. Brill, 2021).

Buell, Paul D., "How did Persian and other western Medical Knowledge Move East, and Chinese west? A Look at the Role of Rashīd al-Dīn and others", *Asian Medicine, Tradition and Modernity*, 3 (2007), 278–294.

Buell, Paul D., "Indochina, Vietnamese Nationalism, and the Mongols", in Volker Rybatzki, Alessandra Pozzi, Peter W. Geier and John R. Krueger (eds.), *The Early Mongols: Language, Culture and History Studies in Honour of Igor de Rachewiltz On the Occasion of His 80th Birthday* (Bloomington: Indiana University Press, Indiana University Uralic and Altaic Series 173, 2009), 21–29.

Buell, Paul D., "The Early Indian Ocean Trade in Medicinals: Some Indications from Chinese Sources", in Angela Schottenhammer (ed.), *Tribute, Trade and Smuggling* (Wiesbaden: Otto Harrasowitz, 2014), 133–140.

Buell, Paul D., "The Sung Resistance Movement, 1276–1279: The End of an Era", *Annals of the Chinese Historical Society of the Pacific Northwest*, III (1985–1986), 138–186.

Buell, Paul D., "Tibetans, Mongols and Cultural Fusion", in Anna Akasoy, Charles Burnett und Ronit Yoeli-Tlalim (eds.), *Islam and Tibet, Interactions along the Musk Route* (Aldershot, Hants: Ashgate, 2010), 189–208.

Carswell, John, *Blue and White: Chinese Porcelain around the World* (Chicago: Art Media Resources, 2000).

Chipman, Leigh *The World of Pharmacy and Pharmacists in Mamlūk Cairo* (Leiden and Boston: E.J. Brill, 2010).

Ciolcîltan, Virgil. *The Mongols and the Black Sea Trade in the Thirteenth and Fourteenth Centuries*, Samuel Willcocks translator (Leiden and Boston: E.J. Brill, 2012).

Dols, Michael (ed. and transl), *Medieval Islamic Medicine* (Berkeley: University of California Press, 1984).

Elgood, Cyril, *A Medical History of Persia and the Eastern Caliphate* (Cambridge: Cambridge University Press, 1951).

Endo Mitusaki 遠藤光曉, "Wang Shuhe 'Mai jue' perushiya goyaku ni hanei shita jūshi seiki sho chūkokuon" 王叔和《脈訣》ペルツャ語譯ニ反映した14世紀初中國音, in Yu Aiqin 余靄芹 and Endo Mitsuaki (eds.), *Hashimoto mantarō kinen chūko gogaku ronshū* 橋本萬太記念中國語學論集 (Tōkyō 東京, 1997), 61–77.

Garrett, Frances, "Critical Methods in Tibetan Medical Histories" *Journal of Asian Studies* 66 (2007), 363–387.

Gutas, Dimitri, *Greek Thought, Arabic Culture, the Graeco-Arabic Translation Movement in Baghdad and Early 'Abbāsid Society (2nd–4th/8th–10th centuries)* (New York: Routledge, 1998).

Klein-Franke, F., and M. Zhu (1998), "Rashid ad-Din and the *Tansuqnamah*, The Earliest Translation of Chinese Medical Literature in the West", *Le Muséon*, 111 (1998), 427–445.

Klein-Franke, F., and M. Zhu, "Rashīd Ad-Dīn as a Transmitter of Chinese Medicine to the West", *Le Muséon*, 109 (1996), 395–404.

Kong S.Y. 江潤祥 (ed.), *Huihui yaofang* 回回藥方 (Hong Kong: Hong Kong 香港, Zhong Bianyi Yinwu Youxian Gongsi 香港編譯印務有限公司, 1996).

Lev, Efraim, "Mediators between Theoretical and Practical Medieval Knowledge: Medical Notebooks from the Cairo Genizah and their Significance", *Medical History* 57: 4 (October 2013), 487–515.

Lev, Efraim, and Zohar Amar, *Practical Materia Medica of the Medieval Eastern Mediterranean According to the Cairo Genizah* (Leiden and Boston: Brill, 2008).

Levey, M., *Early Arabic Pharmacology: An Introduction based on Ancient and Medieval Sources* (Leiden: Brill, 1973).

Martin, Dan, "Greek and Islamic Medicines' Historical Contact with Tibet: A Reassessment in View of Recently Available but Relatively Early Sources on Tibetan Medical Eclecticism", in Anna Akasoy, Charles Burnett und Ronit Yoeli-Tlalim (eds.), *Islam and Tibet, Interactions along the Musk Route* (Aldershot, Hants: Ashgate, 2010), 117–143.

Matsui Dai 松井太, "Mongol Globalism Attested by the Uighur and Mongol Documents", https://www.academia.edu/852303/Mongol_Globalism_Attested_by_the_Uigur_and_Mongol_Documents (accessed April 9).

Monteiro, João Gouveia (ed.), *Aljubarrota Revisitada* (Coimbra: Impresa de Universidade, 2001).

Nasrallah, Nawal, *Annals of the Caliphs' Kitchens, Ibn Sayyār al-Warrāq's Tenth-Century Baghdadi Cookbook, English Translation with Introduction and Glossary* (Leiden and Boston: E.J. Brill, 2007).

Needham, Joseph et al., *Science and Civilization in China* (Cambridge: Cambridge University Press, 1954–2015).

Nutton, Vivian *Ancient Medicine* (London and New York: Routledge, 2007).

Park Hyunhee. *Mapping the Chinese and Islamic Worlds, Cross-cultural Exchange in Premodern Asia* (Cambridge: Cambridge University Press, 2012).

Pormann, Peter E., and Emilie Savage-Smith, *Medieval Islamic Medicine* (Washington, D.C.; Georgetown University Press, 2007).

Pormann, Peter E., *The Oriental Traditions of Paul of Aegina's Pragmatei* (Leiden and Boston: Brill, 2004).

Rosa, Luis, *O Dia de Aljubarrota* (Lisboa: Editorial Presença, 2008).

Sadek, M.M., *The Arabic Materia Medica of Dioscorides* (Quebec: Les Editions du Sphinx, 1983).

Said, Hakim Mohammed (ed., and transl), *Al-Bīrūnī's Book on Pharmacy and Materia Medica* (Karachi: Hamdard National Foundation, 1973).

Song Xian 宋峴, *Huihui yaofang kaoshi* 回回藥方考釋, Vols. 1–11, Beijing 北京: Zhonghua shuju 中華書局 (Zhongwai Jiaotong Shiji Congkan 中外交通史籍叢刊), 2000.

Speziale, Fabrizio, "A 14th Century Revision of the Avicennian and Ayurvedic Humoral Pathology: The Hybrid Model by Šihāb al-Dīn Nāgawrī", *Oriens* 42 (2014), 514–532) (2014a).

Speziale, Fabrizio, "The Persian translation of the *tridoṣa*: lexical analogies and conceptual incongruities", *Asia* 68:3 (2014), 783–796) (2014b).

Speziale, Fabrizio, *Culture persane et médecine ayurvédique en Asie du Sud* (Leiden and Boston: Brill, Lam edition), 2018.

Ullmann, Manfred, *Islamic Medicine* (Edinburg: University Press, 1978).

Unschuld, Paul U., *Medicine in China: A History of Pharmaceutics* (Berkeley: University of California Press, 1986).

Unschuld, Paul U., *Medicine in China: A History of Ideas* (Berkeley: University of California Press, 1985).

Vogel, Hans Ulrich, *Marco Polo Was in China: New Evidence from Currencies, Salts and Revenues* (Leiden, Boston and Köln: Brill, 2012)

Waines, David, *Food Culture and Health in Pre-Modern Islamic Societies* (Leiden and Boston: E.J. Brill, 2011).

Weng Tu-chien, "Ai-hsieh: A Study of his life". PhD dissertation (Harvard: Harvard University, 1938).

Index